高等院校土建学科双语教材（中英文对照）

◆ 风景园林专业 ◆

BASICS

植 物 设 计

DESIGNING
WITH PLANTS

［德］ 雷吉娜·埃伦·韦尔勒　编著
汉斯－约尔格·韦尔勒

齐勇新　　　　　　　译

中国建筑工业出版社

CONTENTS

序

　　植物是构成景观建筑学基础的一种设计元素。无论是花园或是公园里那些充满想象力的组合搭配，还是以广袤的群体呈现出一个总体形象，树、灌木、草本植物和花卉总是能让我们感到身心愉悦。纵观全年，植物仿佛是一位变身大师，它们在不同的生长发展阶段都能形成新的空间形态。但不可否认的是，事实并非总是如此。这要看植物的生存环境是否有利于它们的生长，它们也许会枝繁叶茂，也有可能枯萎凋零，最终留下荒野一片。

　　植物设计并不是只要能用植物组合出色彩绚烂的效果就大功告成了，而是需要我们精通特定土壤、生存环境以及植物的类别和品种等方面的知识，尤其是植物的花期和花色方面的知识。

　　使用植物的目的在于渲染气氛、营造空间、打造景观、美化后院、展示花卉，并且还要让人们认识到植物是"园林艺术"的一个重要"组成部分"。

　　作为设计师，我们不仅要处理那些复杂的任务和问题，还要给各种不同的客户做设计。我们给私人客户设计私家园、为社区或是有轨电车沿线设计绿化、为游戏场地配种植植、营造城市空间以及承接城堡公园或是修道院花园这类的大尺度景观设计等。不论以上哪种情况，有一些原则是在设计和施工中必须予以考虑的。

　　"高等院校土建学科双语教材"系列丛书介绍了景观建筑学的基本原则，为了培养读者的空间设计感觉，本书还提供了一些设计手法上的建议以及解决问题的一些方法，即便是那些缺少设计经验的学生读者，本书也同样有用。

　　书中的各个章节给出了循序渐进的介绍。所有重要的内容都做了逐一说明——从植物的生长环境、功能需求到比例、空间限定、质感以及色彩构成等。书中的图表和图例对观点给予了辅助说明。本书的目的并不是要给出一个普遍适用的公式，而是引导读者对场地的特性及其利用以及最为重要的一点——如何去营造气氛等内容建立认识。但是对于是否需要运用严整规矩的建筑化语汇、是否需要采用品种接近的植物来体现极简主义风格、是否需要营造空间层次，以及是否要去表现多种颜色和香味的植物组合在一起枝繁叶茂的样子等问题，都需要个人结合自己的直觉以及自己对空间的理解来作出最终决策。

编辑：Cornelia Bott

FOREWORD

Plants are a design element that form the basis of landscape architecture. As imaginative compositions in gardens and parks, or building extensive structures and creating an overall impression, trees, bushes, herbaceous plants and flowers never cease to amaze us. Viewed throughout the year, a plant appears as a master of transformations, creating new spatial structures with every new phase of growth and development. Admittedly, it is not always reliable. Depending on whether or not the habitat is favorable, it may flourish or wither away, leaving an unkempt scene behind.

Designing with plants does not simply mean arranging plants in a colorful ensemble. Rather, it requires a thorough knowledge of specific soil and habitat conditions, the various types and varieties of plants, and not least their flowering time and the color of their blossoms.

In using plants, the aim is to create an atmosphere, form spaces, shape landscapes, lay out kitchen gardens, develop floral images and understand plants as a significant "building block" in "garden art."

As planners, we have to deal with complex assignments and issues and plan for a variety of user groups. We design a domestic garden for a private client, provide greenery for a residential area or a tram route, plant up play areas, shape urban spaces, and create imposing layouts such as castle parks and monastery gardens. In all these situations, it is essential to take certain principles into account during planning and realization.

The "Basics" series of books transmits the fundamentals of landscape architecture, suggesting possible design approaches in order to develop a feeling for the space that is to be shaped as well as possible solutions, even if the student reader has limited previous experience of design.

The chapters in this volume offer a step-by-step introduction. All significant aspects are elucidated – from habitat conditions and functional requirements to proportions and space definition, textures and color composition. Illustrations and pictorial examples clarify the argument. The aim is to provide not universal formulae, but an understanding of the specifics of a site, its use and above all, the atmosphere that will be created. But the decision to choose a strict, formal architectural vocabulary, to pursue a minimalist approach involving plant varieties that differ little, to create a spatial hierarchy or to bring out the luxuriance of the different plant types with their colors and scents will always be taken through the interplay of your own spatial diagnosis and your intuition.

Cornelia Bott, Editor

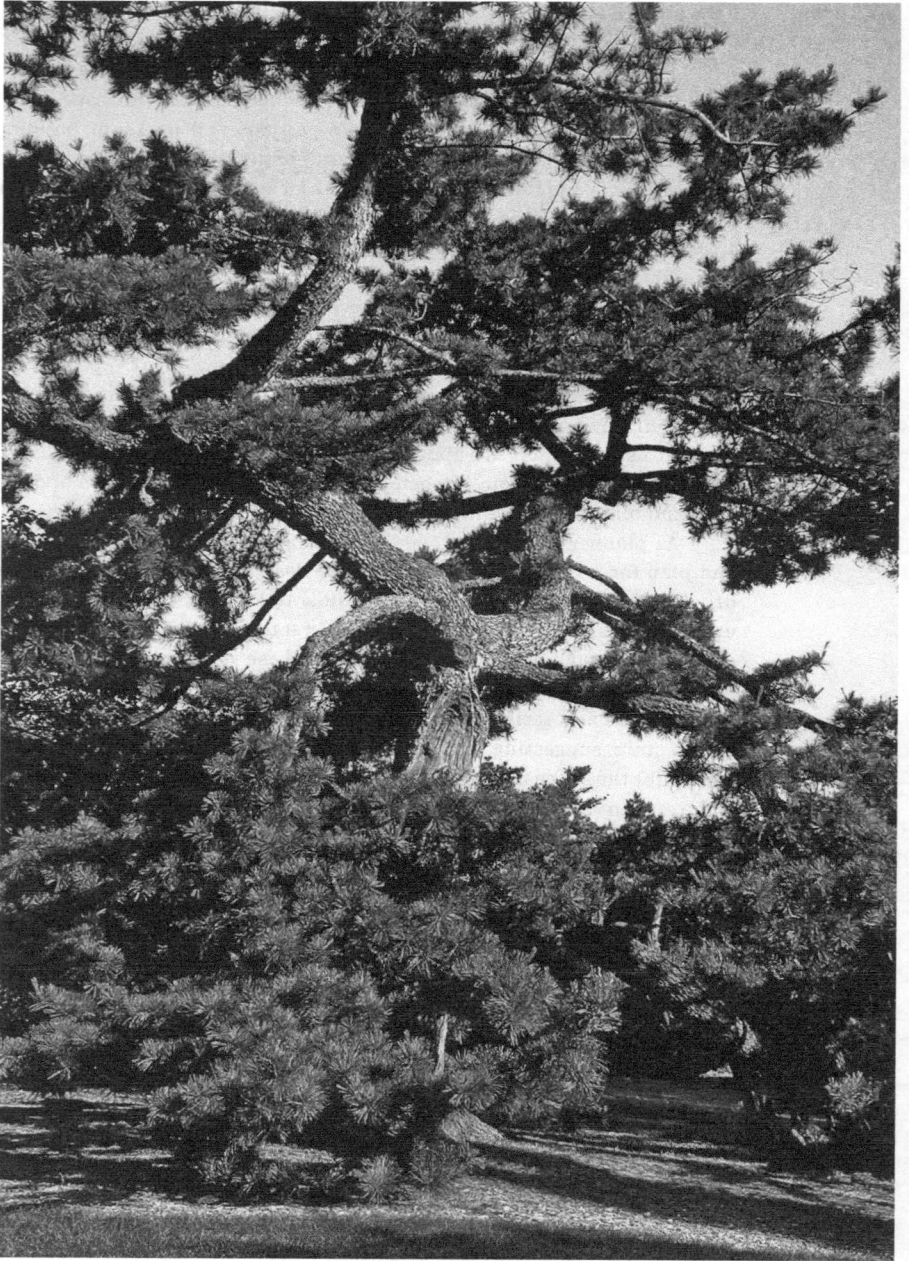

INTRODUCTION

Space and plants – that is, the creation of space using plants in garden and landscape architecture – have always been of exceptional importance in the history of horticulture. However, good design is not exclusively a question of aesthetic feeling. In many ways it depends on the fulfillment of objective contextual requirements. Design using plants demands not only ability, but, to a particularly large degree, knowledge. This means using the plants in such a way that the laws of sensory perception effectively support the purpose behind the design and make it recognizable (the Greek word *aesthesis* refers to the science of sensory perception). This kind of knowledge is made all the more important by the fact that no open-space locality is so like another that prescriptive plans and template images of arrangements of plants can simply be copied. On the other hand, the generally applicable basic principles of design, space creation, order, contrast, balance and repetition can be applied to any site and give good results in plant design.

The plant is a living raw material for shaping open spaces as an important opposing pole and foil to the increasingly technological conditions of our urban civilization. Due to their atmospheric qualities, trees, bushes and herbaceous plants offer a variety of possibilities for open space planning in the gray area between culture and nature. Design using plants embraces very diverse situations, from private houses to the imposing grounds of large buildings to the complex structures of a town's green spaces, with public spaces, pedestrian areas, parks, planted roads into towns, recreational areas, cemeteries and allotments.

This book, *Basics Designing with Plants* will train the eye for design possibilities involving plants in planning open spaces, making it in the process clear that it is only by integrating plants into the planning process that architectural and urban planning blueprints can be broadened into a unified, high-quality concept.

Fig.1:
Climate determines plants' range and distribution.

BASIC DESIGN PRINCIPLES

The text below outlines the key factors that must be recognized and taken into account before undertaking the planning and realization of any objective, regardless of its extent. These factors require certain basic considerations and decisions.

ECOLOGICAL HABITAT CONSIDERATIONS

A journey over open country shows that the faces of the landscape change; in higher and colder regions, temperate fruit and wine growing areas with fertile soil give way to deciduous and coniferous forests with poor subsoil. Plants originating from regions with temperate climate conditions may be sensitive to frost and become easily damaged in winter unless measures are taken to protect them. Plants that grow on fertile soil in their normal habitat will become stunted on poor soil. In designing plantings, it is indispensable to know the natural habitat of the plants as well as their visual effects. › Chapter Plants as a material, Plants: appearance There are a variety of factors that influence the growth of plants: › Figs 1 and 2

› ⏍

_ Climate
_ Soil
_ pH
_ Situation

_ Light
_ Water
_ Nutrients
_ Competition

Climate   Every site is influenced by the macroclimate and its altitude above sea level. These pre-existing natural conditions cannot be changed or circumvented, and determine the distribution of individual plant species. However, the microclimate of a habitat can be influenced. The corners of walls and buildings (for instance in an inner courtyard) create areas protected from the wind, and radiate heat. If further habitat advantages (soil, precipitation etc.) are added, the spectrum of possible plantings becomes wider. The following climate parameters are important for plant growth:

\\ Note:
Ecology is the science of interrelations and interactions between organisms and their adaptation to living conditions. The study of the behavior of individual plant species under environmental influences and the effect of environmental factors on the composition of plant cover is called plant ecology.

11

_ Temperature: winter cold, summer warmth
_ Humidity: summer precipitation, winter precipitation

The most significant factor is the level of cold in winter, as whether a plant species will survive or not depends on the lowest temperature, which is reached during the winter. Frost-hardiness describes the maximum frost temperature that a plant can survive without damage. In the case of summer warmth, it is not the extremes of temperature that are crucial. Rather, it is the sum of the warmth – the average temperature over the summer – that is decisive. Plants need a certain total warmth in order to produce and develop leaves, blossoms and fruits. The milder the climate is, the broader the spectrum of plants that can be used will be.

Soil, situation, pH

Soil is the plant's support. The plant draws water and nutrients from the soil and is anchored in it. Soil structure, water and nutrient content are very important for growth. In choosing species for the habitat that is to be designed, the types of soil present (clay, loam, sand, silt), as well as its pH (degree of acidity or alkalinity), must be taken into account. Different plants require different pH values. The availability of nutrients changes depending on the pH; acid soils are poor in nutrients, while alkaline soils are nutrient-rich.

› 📖

The microclimate will vary according o the gradient and degree of exposure of the ground. South-facing slopes are warmer and drier. North-facing slopes are cooler and moister. On a journey through a hilly region it can be seen that the meadows of north- and south-facing slopes have many distinct types of flowers, and that the composition of the blossom colors vary accordingly.

Light

The light available in a habitat also determines whether a plant can grow and thrive there. Habitats can be divided into those "in full sunlight," "direct sunlight," "out of direct sunlight," "partial shade," and "full shade". There are plants that can only tolerate sunlight or shade, but also plants

📖

\\ Note:
The composition of the soil can be changed.
However, this requires long-term maintenance,
as otherwise the existing natural conditions
will sooner or later reassert themselves.

that will tolerate both, such as the snowberry. Due to growth, the size of the plants and the intervals between them change, so that the availability of light changes over time. Shade will be introduced beneath trees, in particular. › Chapter Plants as a material, Time dynamics Light exposure is not only a factor in the selection of individual plants, but can also influence the overall character of a display or a part of the garden. This is most evident in significantly shaded or sunny habitats. Shaded displays are distinguished mainly by the forms, colors and textures of the leaves of woody and herbaceous plants, as flowering activity is greatly reduced in shady places.

Water is the most important building material and fuel for plants. Under natural conditions, the level of precipitation is therefore of great importance. In particular, summer precipitation protects the plants from drying out under high temperatures and high levels of sunlight during the growing season. In winter, most plants are in a state of rest due to the loss of their leaves, and have less need to replace water. Winter precipitation (snow) is important for frost-sensitive species, as the blanket of snow protects the parts near and below the ground from heavy frost. Frost without snow cover can be damaging to evergreen plants. While water evaporates from their leaves, they cannot extract water from extensively frozen soil and therefore become dehydrated (frost dryness).

Soil's natural water content is determined by precipitation, the level of the local water table, the structure and permeability of the soil, and the gradient of the land. The amount of water available to a plant depends on the structure of the vegetation profile and its breakdown by species. However, plants have very diverse moisture requirements. Some love dry conditions, while there are also species that flourish in water. For instance, pine trees and gorse grow in isolation in permeable sandy soils in bright sunlight together with scrub grass in the natural landscape. They are xerophytic. Their leaves are adapted to harsh environmental conditions and are therefore hard and needle-like, small and linear.

› 𝕚

𝕚
\\ Note:
Carefully aligned water sprinklers can improve habitat conditions for plants, but also increase maintenance costs. Excess water, e.g. on compacted soil, can be reduced using drainage management.

In nature, many plants have similar ecological requirements. In such a competitive situation, weaker species are often displaced into habitats in which the competitors themselves can no longer flourish. Competition arises for instance from thorough overshadowing by tall species, which do not admit enough light for shorter species. In plant-related design, the factor of time should therefore be taken into account. During planting, a garden looks bare, but the plant growth becomes ever stronger over the course of time. The tree crowns in particular cast increasing amounts of shade on the plants beneath them, initiating competition for light, moisture and nutrients. It is therefore important to be aware of the growing behavior of plants including their root systems, growth pattern, and size.

Help in planning the habitat-appropriate use of plants and thereby a harmonious and not overly labor-intensive display of woody plants and herbaceous plants can be provided by "index systems." Woody plants are sorted using a four-digit code. The first digit indicates the habitat:

_ Marsh and swamp
_ River meadow and riverbank woody plants
_ Species-diverse forests and woody plant groups
_ Species-poor forests and woody plant groups
_ Moors and sand-dunes
_ Steppe woody plants and low-moisture forests
_ Woody plants of cool, moist forests
_ Mountain forests and alpine bushes
_ Landscape hedges and decorative plants

The second number indicates the most important soil habitat factors, the third indicates aboveground factors such as light and temperature, and the fourth the size to which the plant grows. These index numbers provide information for habitat-appropriate planting, but are not a plant sociology classification system. Many woody plants can adapt to more than one habitat, resulting in numerous combinations and transitions. Herbaceous plants are also classified by growing zone using four-digit codes, which however are a formulation not only of the ecological habitat, but also of their function. The first digit represents the habitat:

_ Wood
_ Edge of wood
_ Open expanse
_ Bed
_ Rockery
_ Water edge

14

The second digit indicates the selective group (or function), the third indicates habitat requirements and the fourth gives any special indications for use.

USER AND FUNCTIONAL REQUIREMENTS

Plants' habitat requirements determine what plants can be used for the intended site. Plant selection is determined by the practical functions they are intended to fulfill, and their aesthetic and creative qualities. Plant requirements should be ascertained bearing in mind functional requirements during the initial planning phase, and cleared with the client, including the necessary maintenance care. For instance, in the planning of a private garden, the planner should enter into the individual wishes and space creation envisioned by the commissioner to the extent that functional requirements for this plot play a large role in the planning process. If the design assignment is intended for a particular user group, e.g. an open-space design for a hospital, a garden for a housing complex for the elderly, a park or play area, or a cemetery, the demands and requirements of the target group in question should be defined sufficiently in advance and agreed upon with the client.

In planning a play area it should be borne in mind that the trees and bushes should be capable of taking punishment. Children like to run around in bush plantations, and tend to pick leaves, twigs, blossoms and fruits. › Fig. 3 Poisonous plants must therefore not be used for play areas. › Tab. 1 Trees, on the other hand, should be included in the plan, to provide the necessary shade in the summer. › Fig. 4 Fallen branches and

꩜

\\ Example:
A garden with solitaire trees is to be planted with large herbaceous beds. The appropriate index codes would be 4.3.3. 4 = bedding herbaceous plants, 3 = herbaceous plants from mountain forests and high-altitude open sites associated with woods or the edge of woods, 3 = herbaceous plants that like cool habitats and can tolerate periods of shade. The fourth number indicates, among other things, sociability. 4.3.3.7 would be planted individually or in small groups (Anemone japonica), 4.3.3.4 are non-expanding or weakly expanding species, which can be planted in combination (Astilbe x arendsii).

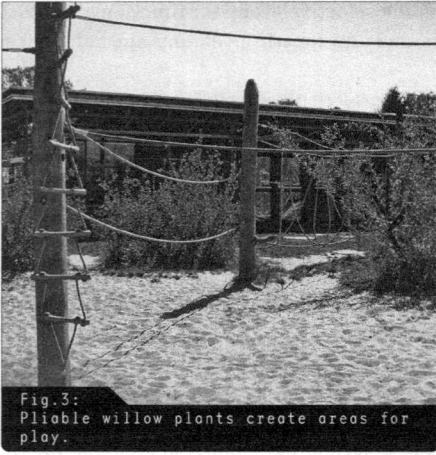

Fig.3:
Pliable willow plants create areas for play.

Fig.4:
In planting play areas, landscapes can be created for children.

twigs of trees and bushes will be used by children as "tools." Older children will clamber on both small and large trees. Young people require spaces for retreat, but they also require areas where they can play games and show off. The planning of a housing complex for the elderly presents entirely different requirements. In this case, the focus is on the experience the planting provides, and on allowing the residents to meet. Colors, shapes and textures should be utilized in a varied and attractive way. Paths should include accompanying shade planting and benches, so that the plants can be viewed in a leisurely way. Seating should be outwardly visible, but be comfortably framed by pergolas and herbaceous plants.

In planning the spatial impact, it is fundamentally important to take into account the larger-scale spatial relationships (lines of sight), the available and planned access to streets and the road network, and the modeling of the terrain at an early stage, and to establish this in agreement with all those involved in the planning process. In large projects, realization in several stages is possible. Intensive design involving plantings and other significant elements such as pergolas, water features, seating and illuminations can take place during a later construction phase.

For instance, the laying out of a cemetery takes place in stages. It generally involves large surface areas, which may not be fully used, depending on density of occupancy. Demanding access routes via road and water networks will only really be required in the first stage of construction. However, the spatial framing for the grounds as a whole (groups of

| Tab.1: | | | |
| Poisonous plants | | | |

Degree of toxicity	Botanical name	English name	Poisonous part of plant*
Highly poisonous	Aconitum (all species and varieties)	Aconite	all
	Daphne (all species and varieties)	Daphne	all
	Taxus (all species and varieties)	Yew	all except false-fruits (arils)
Poisonous	Buxus sempervirens (all varieties)	Boxwood	all
	Convalleria majalis	Lily of the valley	all
	Crocus (all species and varieties)	Crocus	bulb
	Cytisus (all species and varieties)	Broom	seed pods
	Digitalis (all varieties)	Foxglove	all
	Euphorbia (all species and varieties)	Spurge	all, especially sap
	Euonymus (all species and varieties)	Spindle tree	seeds, leaves, bark
	Hedera helix	Ivy	all
	Juniperus (all species and varieties)	Juniper	all, especially the tips of twigs
	Laburnum (all species and varieties)	Laburnum	all, especially blossoms, twigs, roots
	Lupinus (all varieties)	Lupin	seeds
	Lycium halimifolium	Chinese wolfberry	all
	Rhododendron (all species and varieties)	Rhododendron	all
	Robinia pseudoacacia	Black locust tree (false acacia)	bark
	Solanum dulcamara	Bittersweet or woody nightshade	berries in particular
Slightly poisonous	Aesculus (all species and varieties)	Horse chestnut (or buckeye)	unripe fruits and fruit cases
	Fagus sylvatica	European beech	beechnuts
	Ilex (all species and varieties)	Holly	berries
	Ligustrum (all species and varieties)	Privet	fruits
	Lonicera (all species and varieties)	Honeysuckle	fruits
	Sambucus (all species and varieties)	Elder	everything except the ripe fruits
	Sorbus aucuparia	Rowan	fruits
	Symphoricarpos (all species and varieties)	Snowberry or waxberry	fruits
	Viburnum (all species and varieties)	Viburnum	fruits

* slightly poisonous parts of plants still cause severe complaints

trees and border plantings) should be created during this period, so that after completion of all surfaces, individual sections' signs of construction can barely be seen. This means that all framing displays will show the same stage of development and growth. This creates the impression of a unified overall layout. This way of proceeding can also be applied to other layouts (residential areas, leisure parks and sports facilities) and should always be utilized to achieve a harmonious overall effect. › Chapter Plants as a material, Time dynamics

RELATIONSHIP TO LOCATION

Any area design is created in its own specific context. The site and its surroundings are one part of this, but community and socio-cultural conditions also play an important role. In developing a design solution, an intensive engagement with the site, its surroundings, its history and its users is very helpful. During analysis, the systematics, dependencies and relationships between the elements of the site, among other things, will be worked out in advance. They form the fundamental structure, the basis for the design. A design can either integrate harmoniously with this structure, or interpret it with an alternative approach. Equally, a consciously opposing position may be sought, or a design approach independent from such structures developed.

Concerning oneself with the site contributes to an understanding of the particular circumstances that influence the situation on the site, and to integrating them into the design process.

› 🛈

Landscape and urban planning issues

The significant basis for design applied to open space is the topography of the site to be landscaped. Terrain, whether completely flat, inclined, terraced, containing a variety of gradations or undulating, always has implications for the creation of space and the relationship between indoors and outdoors. If the plot incorporates extensive views of the surrounding landscape, research should be carried out on which alignments between plot and landscape or within the landscape could be interesting. › Fig. 5 In

🛈
\\ Note:
Further information on the subject of drafting in context can be found in *Basics Design Ideas* by Bert Bielefeld and Sebastian El khouli, Birkhäuser Verlag, Basel 2007.

🛈
\\ Note:
In the re-planning of architecture and open space, it is important that architects, town planners and landscape architects work together closely to develop an interrelated concept (see Figs 6 and 7).

Fig.5:
Fruit tree orchards mark a landscape.

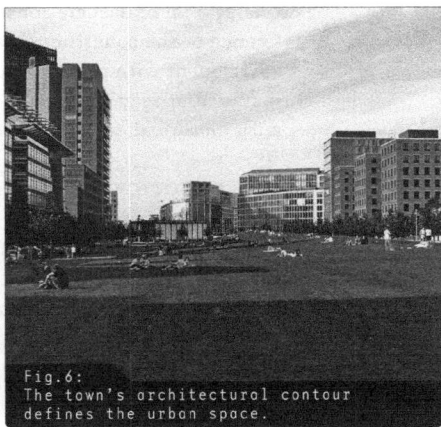

Fig.6:
The town's architectural contour
defines the urban space.

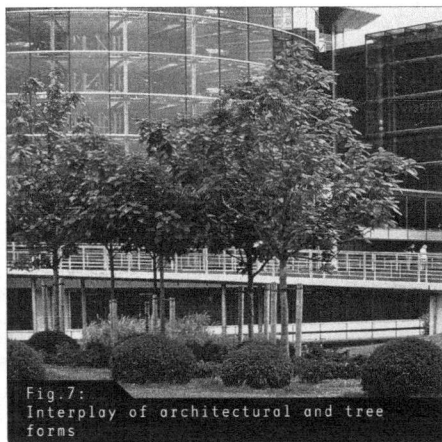

Fig.7:
Interplay of architectural and tree
forms

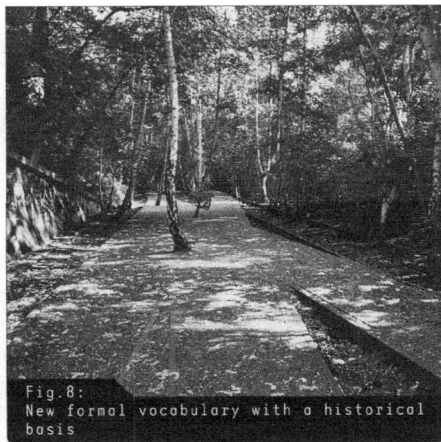

Fig.8:
New formal vocabulary with a historical
basis

Historical
issues

an environment marked by human activity, influences from civilization are
as important for the design as natural ones. Many buildings, streets and
trees serve as reference points for open-space planning.

Concern with the character of the site is not restricted to directly spatial matters. Every alternative is also a reaction to the history of the site,
while also shaping the future. Shaping and altering an existing situation
is an intervention, which is inevitably perceived by the environment as
part of a continuous process. However, it should be borne in mind that

this system of associations will always be on the same level as the significance of the construction assignment. This means that in a construction assignment with great social significance, a park layout or a memorial, for instance, it may be proper and appropriate to thematize the association with historical events. › Fig. 8

FUNCTION

Plants have a variety of properties and therefore a variety of functions and implications for the environment and for human beings. › Fig. 9 For human beings, the most important question is generally whether and how the plants can fulfill economic and technical functions and the effect

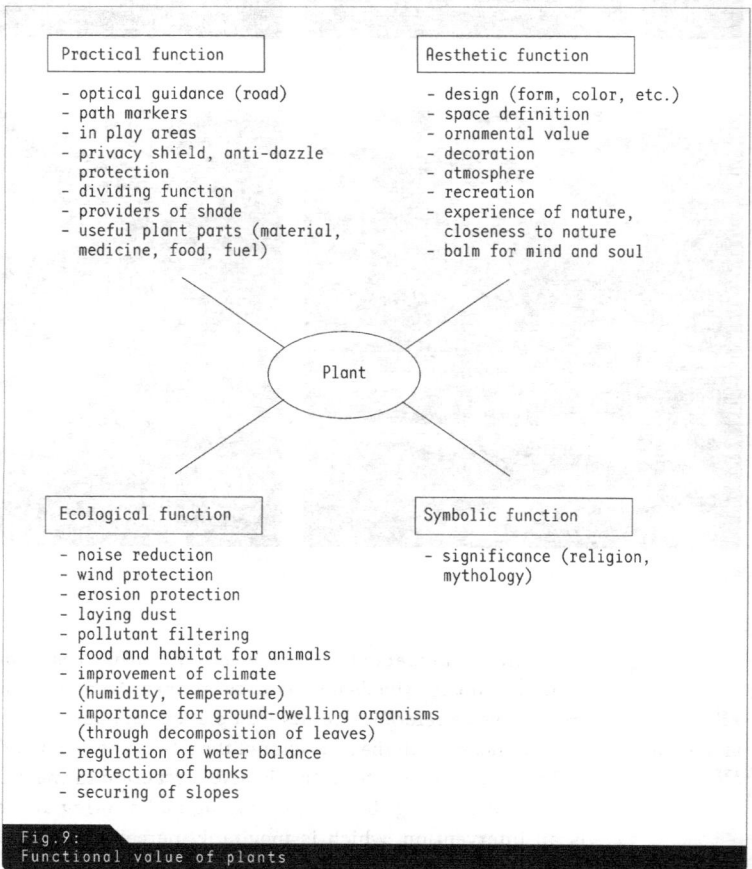

Practical function	Aesthetic function

- optical guidance (road)
- path markers
- in play areas
- privacy shield, anti-dazzle
  protection
- dividing function
- providers of shade
- useful plant parts (material,
  medicine, food, fuel)
	- design (form, color, etc.)
	- space definition
	- ornamental value
	- decoration
	- atmosphere
	- recreation
	- experience of nature,
	  closeness to nature
	- balm for mind and soul

Plant

Ecological function	Symbolic function

- noise reduction
- wind protection
- erosion protection
- laying dust
- pollutant filtering
- food and habitat for animals
- improvement of climate
  (humidity, temperature)
- importance for ground-dwelling organisms
  (through decomposition of leaves)
- regulation of water balance
- protection of banks
- securing of slopes
	- significance (religion,
	  mythology)

Fig. 9:
Functional value of plants

Fig.10:
A group of trees with a distinctive
cross has a path-marking function.

of their appearance (form, color etc.). The appearance of plants (character-istics, foliage, flowers, fruit) has a very high experiential value for people, and its importance to the mind and soul should not be underestimated. Plants are of elementary importance to ecology and the climate. Aesthetic, ecological and technical functions need not be mutually exclusive, but may work together.

Function in
space creation

Open spaces are largely structured by plants. To be precise, they create levels and demarcate different altitudes (from the tree to the flow-ering bulb). Groups of plants or solitaire woody plants can also create a connection between different functional spaces. The size and form of spaces can be structured by groups or series of plants. > Chapter Spatial structures

Path signing

Plants have a path-signing function as path markers, landmarks and necessary marking (for instance, to mark the edge of a slope). > Fig. 10 In the case of paths and roads, hedges, individual groups of woody plants or larger stands of trees may give appropriate visual guidance.

Protective
function

Plants can be used effectively in a variety of ways to protect against climatic or environment-hostile influences (e.g. noise or gales). Extensive tree crowns protect people from bright light and heat in summer. In winter, the bare branch structures allow a clear view and let sunlight through. Thick, bushy hedges may minimize or exclude wind, noise or dust entirely, as required. On sloping ground or embankments, plants growing low to the ground may provide protection against erosion.

21

Fig.11:
Groups of trees create spaces for
retreat and create atmosphere.

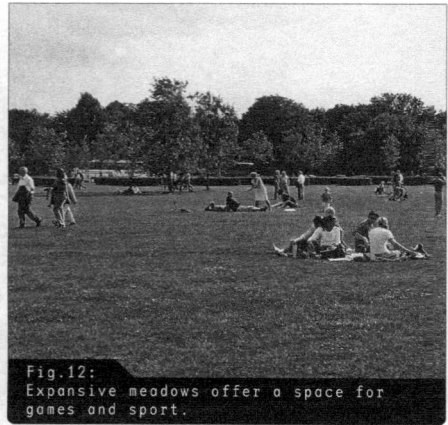

Fig.12:
Expansive meadows offer a space for
games and sport.

Plants often perform different functions simultaneously. For instance, clipped hedges or free-growing woody plants frame a car park's parking slots. In structuring the rows of parking spaces, their effect is simulta- neously space-creating and protective. Trees roof over a car park with their extensive canopies and provide protection from sunlight in summer.

Atmosphere and recreation

Most people only develop an interest in garden design when setting out to plan their own gardens. A garden should fulfill its intended func- tion with a formally balanced design. By awakening certain moods in those contemplating it, it gains a particular effect and aura. This may be a feeling of peace, or of leisure, relaxation, security or seclusion. People experience joy through contemplating plants. They seek feelings of wellbeing in places with balance and surprises. Gardens and parks are formally a compro- mised idealized image of the world. They awaken a suggestive power in people, an endeavor to draw closer to the first garden, or Eden. At the same time, gardens and parks are always a reflection of their time period. Cur- rent social, design, economic, ecological and functional conditions should be made visible. › Figs 11 and 12 Entirely different gardens, urban spaces and squares may be created from one and the same formal design, depend- ing on the priorities applied when arranging the plants. For instance, the city is characterized by its heterogeneity, its simultaneity. The landscape architecture of the city springs from diversity and responds to the re- spective qualities and poetry of different places. › Fig. 13 The appearance of plants plays a significant role in this and may give a square, a garden or a park layout a number of different characteristic features. › Tab. 2 For instance, free-growing woody plants give a layout a natural, landscape-like

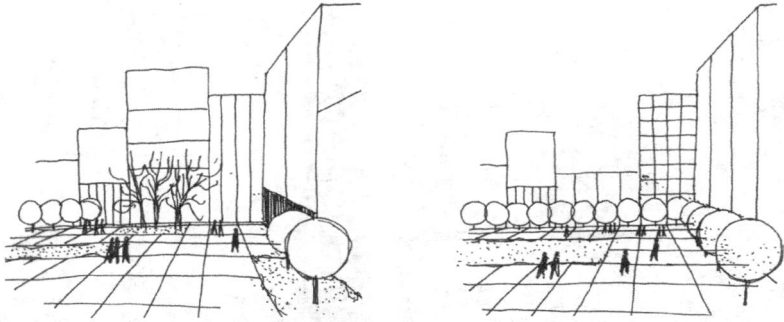

character, while formally clipped woody plants provide an imposing, formal appearance. Different coloration, textures and structures create different overall images and moods. › Chapter Plants as a material, Plants: appearance In choosing suitable plants, the lie of the terrain, the soil texture, climatic conditions, and the estimated amount of care required play a decisive role. It is also important to remember that every garden, regardless of its size, requires a certain amount of discipline and attention in order for the plan to remain recognizable years later and not be literally overgrown by fast-growing, too densely or randomly placed plants. Whatever style the design is based on, throughout the years the atmosphere in the garden will always carry the personal signature of the "gardener." And this is how it should be, as it is only in this way that a garden can unfold its own particular atmosphere and seclusion.

Orientation and guidance

A planned route is characterized by a visible and directly reachable goal. It is the most natural form of guidance, and the one most natural to human motion. The more natural and refined the way in which the intervening spaces are integrated, the more interesting the route proves to be. The user's "instinctive compulsion" to reach a selected goal is encouraged. However, the goal should not become visible too soon, in order to counteract the desire to cut straight across to it. Plants, seating and viewing points may act as indicators, as way markers, landmarks or spatial marks (e.g. the border of a meadow). In the case of paths and streets, bushes, individual trees and groups of woody plants or larger stands of trees may support visual guidance. › Fig. 14 In particular, rows of trees can indicate a direction from a long way away. › Chapter Spatial structures, Grouping

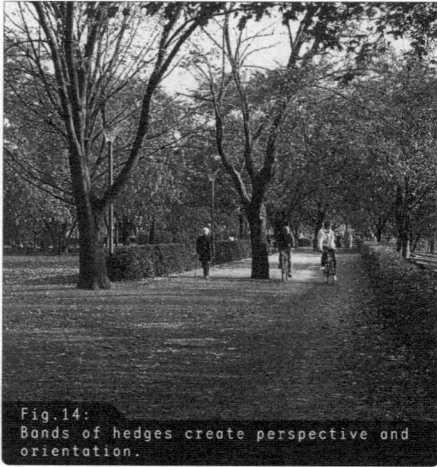

Fig.14:
Bands of hedges create perspective and orientation.

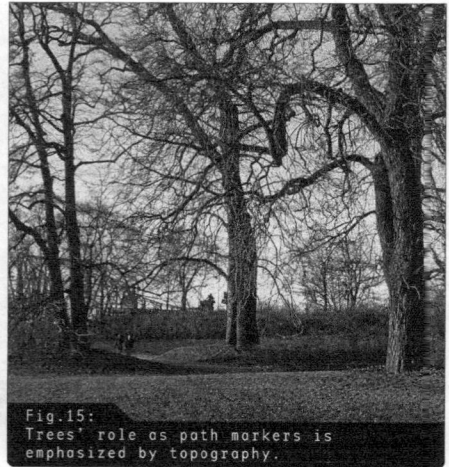

Fig.15:
Trees' role as path markers is emphasized by topography.

Tab.2:
Characters of plants

light	/	dark
imposing	/	modest
quiet	/	loud
exuberant	/	meager
strict	/	loose
extensive	/	intensive
formal	/	scenic
varied	/	monotonous
natural	/	artificial
expansive	/	intricate
robust	/	sensitive
monumental	/	delicate

In designing curved guidance routes, care must be taken that they do not become an end in themselves. Every curve should be the result of actual existing topographical or scenic factors (modeling, plants, a good view). › Fig. 15

SPATIAL STRUCTURES

In open-space planning, as in architecture, the focus is on the creation of spaces. Human beings need and seek these places, as locations where they can orient themselves and in some respects find protection. Demarcation, for which a number of different concrete possibilities such as ground modeling, plants or constructed elements exist in an open-space setting, always plays a role in space creation. The sense of space can be awakened by minimal indications, for instance a ditch or a bush, a dip or a low-hanging tree crown. When, in the course of creating a new layout, the need for a new spatial form and definition in the context of the surrounding objects arises, research into the history of the area can provide useful indications. A space may have a geometric ground plan, but this is not essential; the more a space is created to be independent from its surroundings, local situation and function, the more freedom one has in choosing the form. In developing spatial structures, an opposition is often set up between poles such as broad and narrow, or near and distant. The basis of this treatment is the perception of expansiveness and comprehensibility, or seclusion and openness.

DEFINING SPACE

Space within a landscape is defined by vertical contours, that is, by lateral demarcation. Pillar-like elements indicating a border may be sufficient. › Fig. 16 In cities, trees and buildings define space. When two buildings are connected by a line of trees, these two elements form a border line. When four such lines enclose a space, a closed tree space is created. If buildings are added to these and a purpose assigned to the space within, a square is created, which may have a variety of public functions. It may serve as a meeting place or contain a market. However, a space may also be defined by a depression within an even expanse, or by creating terraces on a slope by removing earth. In open space, simple indications of bordering and horizontal elements are often sufficient. › Fig. 17 It is important to emphasize the contour in the places where the form of the space is to be made recognizable. › Fig. 18 If they are to make the form of the space clear, the contour's corners and curves must be recognizable › Fig. 19 If spaces are intended for people to linger in for long periods, they should be designed and equipped accordingly. Even a single tree or a pergola overhung with vines, which with use will increasingly assume the character of an arbor, or a pavilion, may fulfill this function. A space in the open can almost always be seen into, as it is usually intended to offer a view. Only a few elements, plants and supports, are needed to provide a spatial limit, an outline of the space. Balanced solutions for details enable an individual

Fig.16:
Pillar-shaped trees border a space.

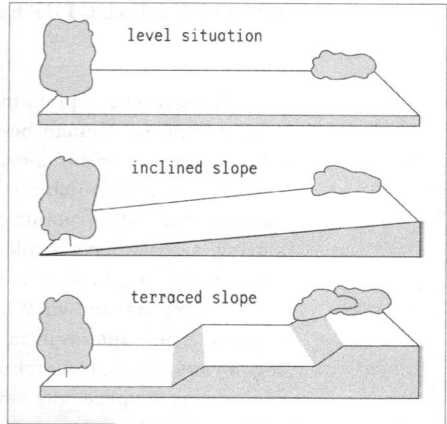

level situation

inclined slope

terraced slope

Fig.17:
Transition from surfaces to space

Fig.18:
Emphasizing the space's contour by means of a curved hedge

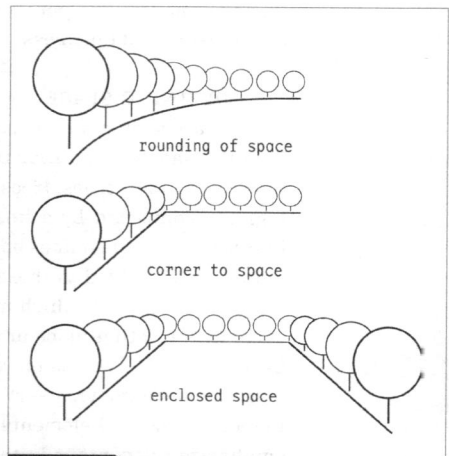

rounding of space

corner to space

enclosed space

Fig.19:
Defining space using rows of trees

atmosphere. > Chapter Basic design principles, Function Groups of trees or solitaire woody plants may form dividing or connecting elements within individual functional spaces. > Fig. 20 By the same token, closed borders to a space can

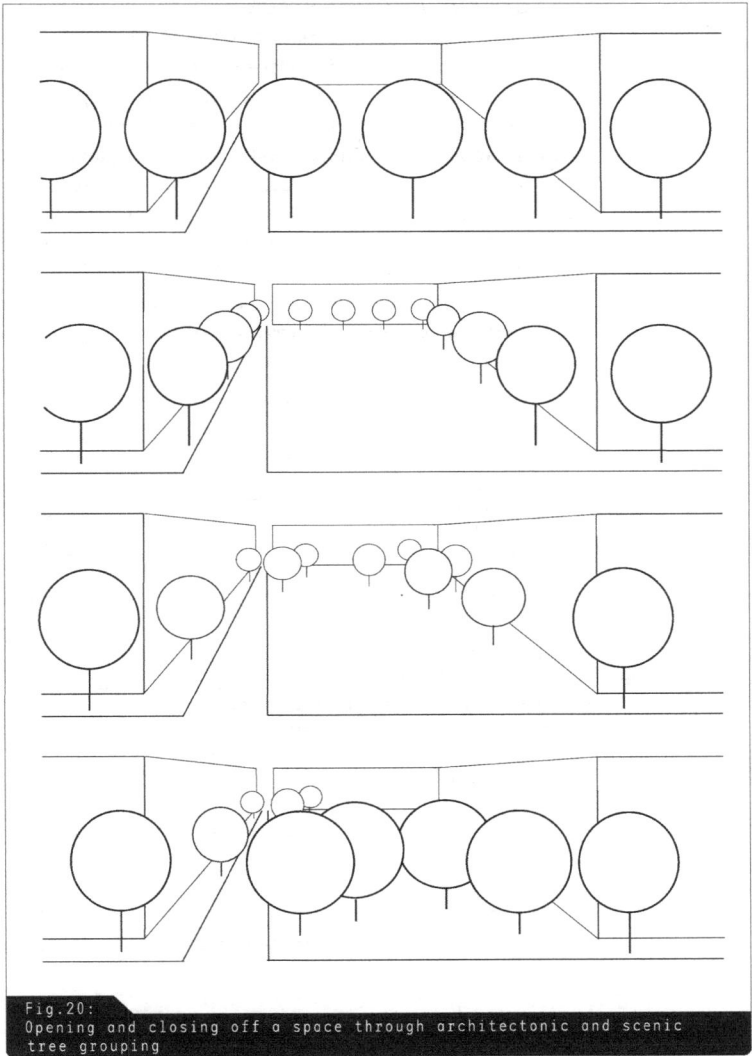

Fig.20:
Opening and closing off a space through architectonic and scenic tree grouping

be opened up by features that interrupt them. In the case of paths and streets, alleys, isolated groups of woody plants and larger stands of trees offer space-defining visual guidance.

ORGANIZATION OF SPACE

To design using plants is to organize. It is only with the formation of intelligible spatial and surface structures that the intention of the designer can be understood, and the functions of open spaces be conveyed. However, the aim of order in design is not monotony, as something homogenous cannot be organized. Rather, an abundance of plants requires a comprehensible order. Above all, design with plants requires the right arrangement. This involves utilizing plants so that their effects do not clash or neutralize each other, but instead heighten one another. This requires individual plant species to be ranked according to their growing behavior, as well as a conscious correlation of plant forms and colors. > Chapters Plants as a material, Principles of design, and Plants as a material, Plants: appearance The ordering of vertical space is a significant design objective in planted installations.

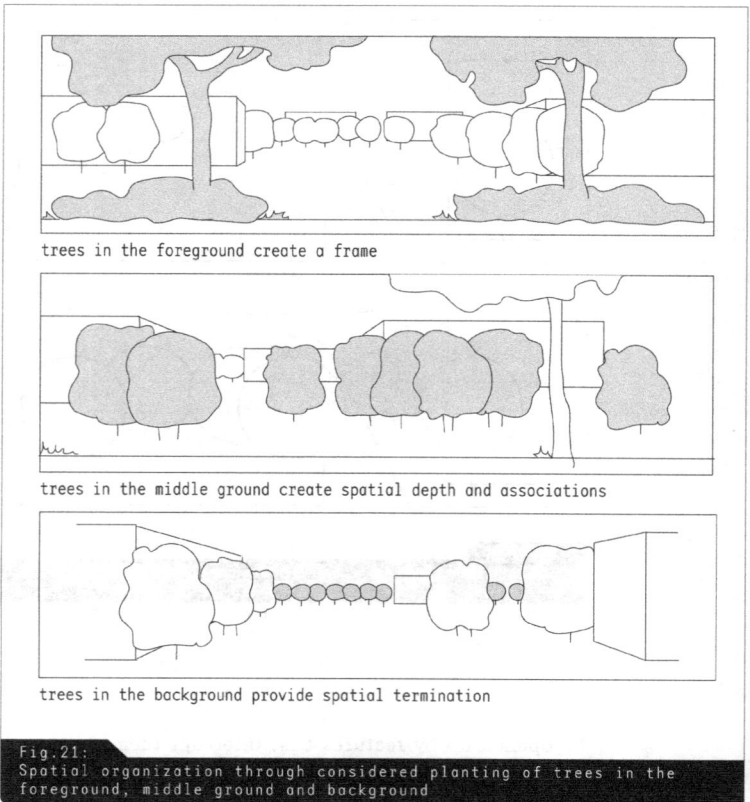

trees in the foreground create a frame

trees in the middle ground create spatial depth and associations

trees in the background provide spatial termination

Fig.21:
Spatial organization through considered planting of trees in the foreground, middle ground and background

Fig.22:
Spatial division and segmentation through low mounds and individual trees

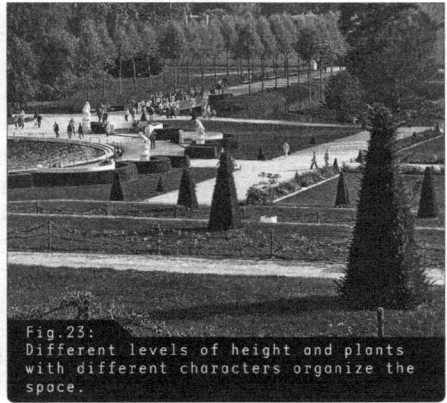

Fig.23:
Different levels of height and plants with different characters organize the space.

The juxtaposition of a tall tree with a display low to the ground assures visual interest. Care must be taken that the plants can coexist in the long term, bearing in mind their individual light requirements and the fact that they will develop to different heights. Competition between them should be avoided as far as possible.

Foreground, middle ground, background

In a visually lucid open space, the spatial organization is reflected in the conscious creation of a foreground, middle ground and background using plants, taking into account size and color relationships that change with distance. Trees in the foreground play a different role from those in the background. Trees in the foreground introduce the architectonic image. They provide the observer with shadows to allow restful contemplation of the view. The trees in the mid-ground provide proportion and create spatial depth. Trees in the background function as spatial demarcation. Background displays have the important task of creatively unifying the garden space. To be effective, a complicated foreground display, a parterre or herbaceous display for instance, requires a calm spatial termination. Background displays often achieve two functions: providing a background and a unity and visual continuity with the surrounding landscape or the neighborhood. › Fig. 21

A succession of areas within the space (open-space sequence), each with a different, assigned purpose and design (e.g. themed gardens), is also an expression of spatial organization. › Fig. 22 One fundamental concern is that for open spaces laid out with plants, unlike architectural spaces, the proportion-giving elements are subject to change due to growth. The space's proportions can best be fixed through regular clipping, which is indispensable in garden situations with an architectural character. › Fig. 23

BORDERS

A space can be bordered using a number of diverse means. Buildings, walls, fences, hedges or ground modeling may present unified physical boundary walls. Composite borders are created by rows of diverse elements along a borderline, such as individual trees, solitaire bushes, constructions with climbing plants, other features, stones, screen walls, and single hillocks. If the space is bordered with plants, then the principle mentioned earlier applies: they change with time. They grow. In isolation, their growing behavior changes. This means that a garden that is broad and open in the years immediately following planting may be an uncomfortably confining space after 20–30 years, usually due to a lack of care and pruning. When choosing plants and establishing the habitat during planning, it is therefore important to know the maximum height of the species involved and their expected behavior over the course of time and depending on situation (e.g. a solitary situation or dense tree or bush groups). Yearly attention, e.g. the thinning out of woody plants, can prevent excessive development and barren lower regions in free-growing hedges and bushes. The height of a border together with the size of the enclosed expanse determines the section of the sky visible, thereby giving the impression of spatial expansiveness or narrowness. A 2 m high hedge loses its spatial impact accordingly as its distance from the observer increases, so that as the size of the enclosed space increases, correspondingly high borders are required. Ground features with a spatially structuring function (ditches, ridges and terraces) demonstrate this relationship even more clearly. Spatial bordering elements either above or below the viewer's eye level (about 1.7 m) should be provided. › Fig. 24 Low surrounds like hedges, trellises, steps and kerbs, if they run crosswise to the line of sight, also contribute to the structuring of open space. They are sufficient to mark garden borders or to differentiate areas with different functions. At the same time, they create a relationship with the surroundings. The garden space is visually expanded by these structuring elements. Clipped walls of trees or free-growing rows of trees, on the other hand, form frames and clearly demarcate spaces. A wide range of space-defining forms of vegetation and elements incorporating climbing plants are available, depending on the desired texture and transparency. The choice ranges from fine-limbed, transparent frameworks of tendrils with light growth to hedges cut to form a massive-appearing wall, or planted walls. › Figs 25 and 26 Overhanging woody plants, bushy herbaceous plants, and trees with sideshoots reaching to the ground obstruct the view, whereas tall trees or large bushes with high branches let the eye roam freely. Winter creates an entirely new spatial effect, with the leaves falling from trees and bushes coming into play. In its contrast of light and dark, color and texture, the

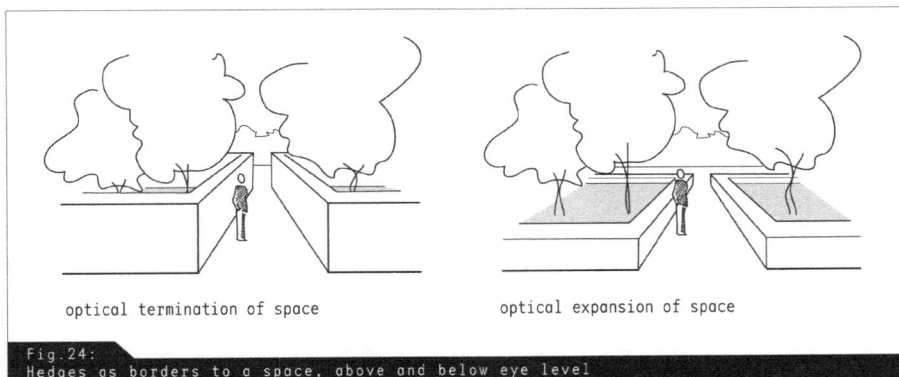

optical termination of space

optical expansion of space

Fig.24:
Hedges as borders to a space, above and below eye level

Fig.25:
Form-clipped hedges create a spatial contour.

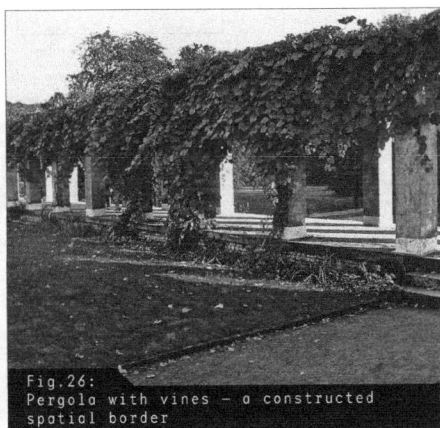

Fig.26:
Pergola with vines – a constructed spatial border

selection of plants should correspond to the overall character of the layout involved (play area, imposing architecture, cemetery etc.). › Chapter **Plants as a material, Plants: appearance**

GROUPING

A connection between separate visual elements is created when we combine them on the basis of similarities to form units or groupings. For example, the arrangement of trees as a group creates a spatial situation. They may be arranged according to a number of different principles, strictly and regularly as a tree grid, or in an irregular and relaxed way as an airy grove.

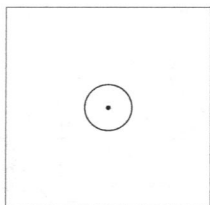 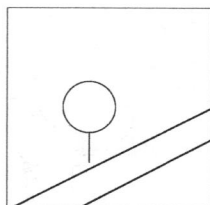

single tree

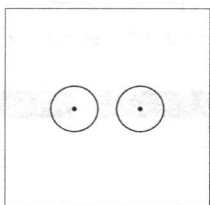 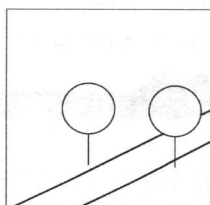

pair of trees

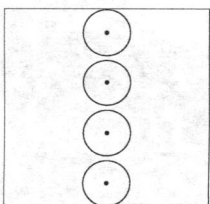 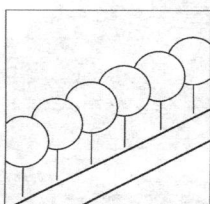

row of trees

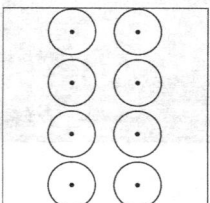 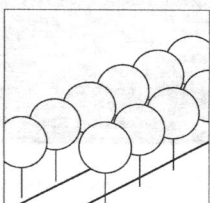

avenue

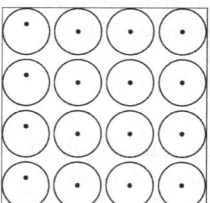 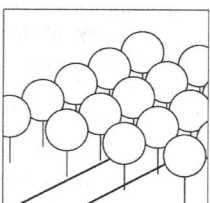

block of trees

Fig.27:
Regular grouping of trees

32

Fig.28:
Solitaire tree

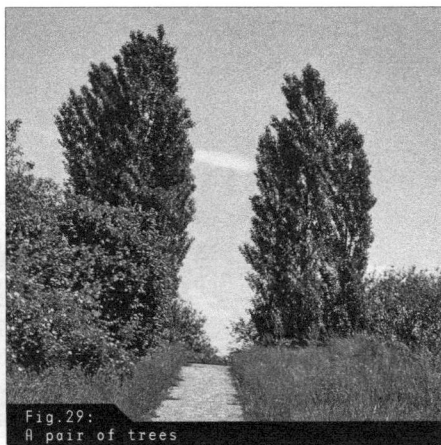

Fig.29:
A pair of trees

Regular tree grouping

Regular, formal groups of trees can create a powerful gesture in urban open spaces. If more than a dozen trees are planted in the same pattern, it is known as a grid planting rather than a group of trees. This is a very simple yet effective design element. > Fig. 27

Solitaire trees

Mature solitaire trees have a powerful effect in a landscape. They are landmarks visible at some distance. In garden design, solitaire trees are either clearly integrated into the overall plan, standing in significant positions – at the end of a path or a sightline, as a central point, or as corner marker of a garden space – or, in order to create a contrast, deliberately placed outside the framework of such an organization. > Fig. 28

Pairs of trees

Pairs of trees are also design elements in the landscape, in gardens and in urban space. > Fig. 29 In gardens, entrances, seating, garden houses, or the transition from one garden to another are often flanked and emphasized by pairs of trees. In urban or architectural contexts, pairs of trees often mark imposing entryways.

Rows of trees

In many European cultural landscapes, trees are among the most important structural elements of the land. Rows of trees are also a recurring design element in horticulture. They define space and impose a rhythm. > Fig. 30 In cities, the courses of rivers, streets and edges of squares are lined with rows of trees. Their ability to give a shape is often much stronger than that emanating from the edges of buildings. A consistent plan for green areas is therefore very important for the creation of a harmonious overall urban image. Rows of trees fulfill a variety of design aspects:

33

Fig.30:
A row of trees

_ They can indicate directions.
_ They can limit views.
_ They can create spaces and linear space.
_ They can harmonize street facades.

Trees can be used as a visually regulating factor if buildings are disparate, or if the overall impression of a street is restless and irregular. Conversely, trees may enliven a monotonous-looking street. › Fig. 31

Avenues

Multiple rows of trees create avenues. Avenues are among the most impressive design elements involving trees. The German word for avenue, *Allee*, comes from the French word *aller* ("to go") and describes a path with trees on both sides. › Fig. 32 In a town, people can stroll or play under trees. Incorporating trees as borders to streets and central plantings means that large expanses of streets can run under rows of trees. Some great boulevards, such as Unter den Linden in central Berlin, have become famous worldwide. This kind of street planting makes the appearance of the town and the individual streets gentler and more artistic. The cultural history of avenues began in the Renaissance, reaching its high point in the 18th century. During the age of absolutism, dead-straight avenues many kilometers long became an expression of human mastery over the landscape. One example is the Route Napoleon with its avenues of poplars. Monarchs and regional rulers had the roads to their castles, country seats and hunting lodges planted with shady trees on either side. For avenue planting, distances of 5–15 m between trees, depending on the type of tree, are recommended. The closer together they are planted, the more pronounced the space-defining effect is.

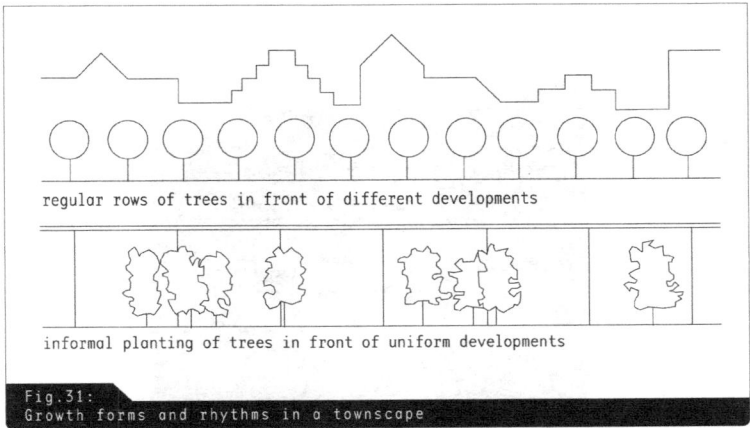

regular rows of trees in front of different developments

informal planting of trees in front of uniform developments

Fig.31:
Growth forms and rhythms in a townscape

Fig.32:
An avenue

Tree block

A tree block is an assembly of trees of the same age and type. They are arranged at regular intervals in every direction, usually on a rectangular expanse of ground. The architectural effect of this regular arrangement is reinforced by using thick-crowned deciduous tree types, e.g. horse chestnuts and maples. In the case of box-cut lindens, a cubic cut for the tree crowns gives this arrangement an almost "built" character. In an urban context, the tree block becomes part of the architecture. Multiple tree block rows running in the same direction create effective axes (a boulevard). Surrounded by comparatively free-growing woody plants (a landscaped park),

Fig.33:
A tree-lined square

a regular tree block suggests an architectonic highlight (central buildings and formal squares).

Tree-lined squares
When more than a dozen trees are planted in the same arrangement, this is no longer a group of trees, but a grid planting. › Fig. 33 However, groups of trees in a city do not only function as spatial elements. They have other important functions, including:

_ The square as a meeting place
_ The square as provider of shade
_ The square as advertising space
_ The square as an area that improves the microclimate

Tree-lined squares are popular because visitors can choose between sun and shade. In the summer months in particular, the shade is appreciated in the afternoon and early evening. The square may be improved by a fountain, colorful flowerbeds, hedges, bushes and walls, which pattern the square's layout.

Scenic tree grouping
Urban design using trees has a dual aesthetic function. It alters the face of the city, and it signifies an influx of nature into the city in that it is not merely a "remnant" of nature. › Fig. 34 Its most important function consists of emphasizing its difference from the artificial, i.e. the buildings. Over the centuries, towns have been built to geometrical patterns, involving grids and strict regular forms. Within this rigid structural framework, landscape tree groupings represent a piece of nature that runs

36

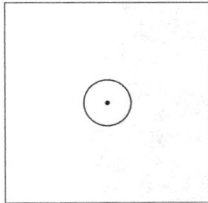

single tree

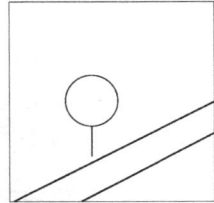

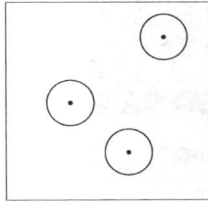

several
solitaires

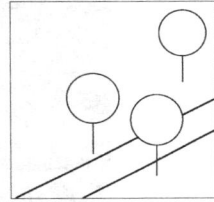

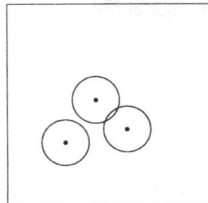

group of three

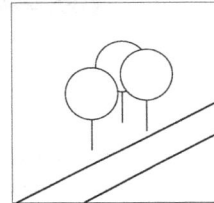

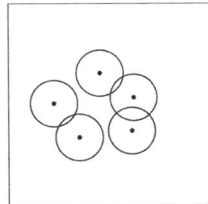

group of five

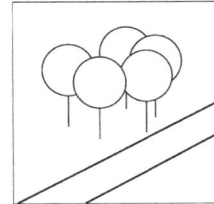

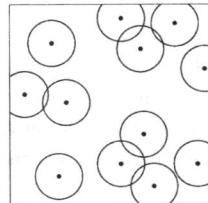

irrgular grove

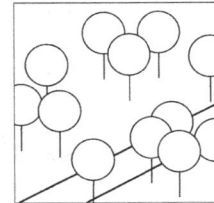

Fig.34:
Scenic grouping of trees

Fig.35:
Group of trees

diametrically counter to the code of rational urban development. Scenic
grouping of trees incorporates a variety of elements:

_ The single tree
_ Several single trees
_ Groups of three
_ Groups of five
_ Grove

Tree groups

Groups of trees have a different function from that of rows of trees
in an avenue. Not only can they emphasize buildings, they can also border
small areas or be a complementary intermediate part of a town's layout.
Freely combined groups of trees offer the designer the possibility of guid-
ing the viewer's eye and giving a further depth to the open space by means
of appropriate arrangement. › Fig. 35 Through the interplay of topography
and the apposite placing of tree groups, charming garden landscapes can
be created. A look at horticulture throughout the ages and across the range
of styles shows that the tree group as a design element has always been
used, in freely arranged and in geometrically coherent forms. In geometrically arranged groups, trees are often planted close together, at intervals
of 1.5–2.5 m, and are therefore described as tree packages.

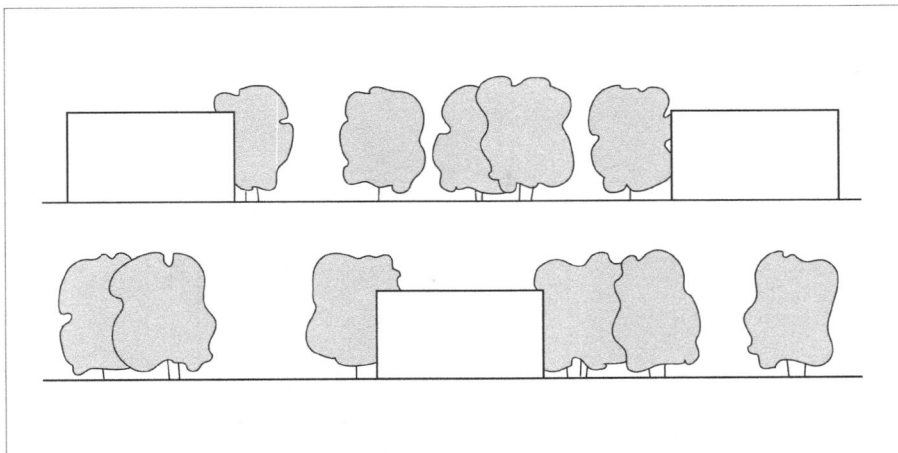

Fig. 36:
Scenically arranged groups of trees can create a visual connection between several buildings or frame a single building.

A few trees or loose tree groupings may be used to frame a building or create a visual connection between several buildings. › Fig. 36 Trees create spatial depth and tone down buildings' hard edges. Buildings of different shapes can also be visually harmonized using a few trees or small groups of trees. › Fig. 37 If a few trees, all of the same kind, are included in diverse developments or housing estates, the impression of a coherent framework is created. This shows how a tree theme can make a town district visible and comprehensible to residents and visitors.

Tree grove

Groves of trees have a variety of characters and evoke a variety of moods. This primarily depends on the choice of tree type (e.g. easily permeable to light or shady, dark or light green leaves, lustrous or matte leaves), but it also depends on how closely the trees are positioned and the planting structure (strict, formal, free, irregular). A grove is created by the planting of woody plants of the same type and age. The open character of a grove can be heightened by the use of loose-crowned, fine-textured tree types (e.g. birch, larch, pine, locust tree) and by a lower level of ground-covering vegetation, meadow or lawn. › Fig. 38 Depending on the species of tree, their age and the intervals between them, an entirely different atmosphere is created. For instance, a light grove of tall beeches is bright and friendly, while a pine grove, with its evergreen coating of needles, is comparatively dark. In contrast to a strictly regular grove, a

Fig.37:
Using trees to unify different styles of architecture

free, naturalistically arranged grove is not subject to any regular grid. The intervals between the trees are created by "scatter planting." The sequence of light and shadowed areas is irregular. Large gaps alternate with small ones, and loose sections with dense. A free grove may give a variety of impressions. Depending on the tree type chosen, the feeling evoked may be Arcadian or melancholic.

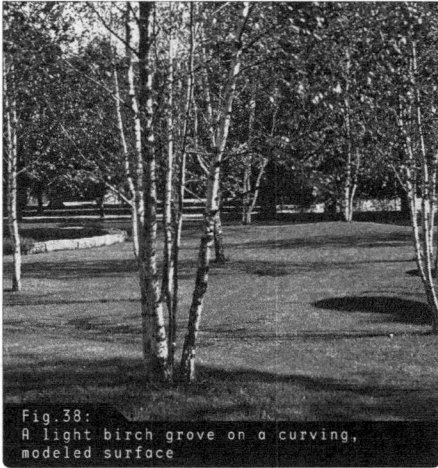

Fig.38:
A light birch grove on a curving,
modeled surface

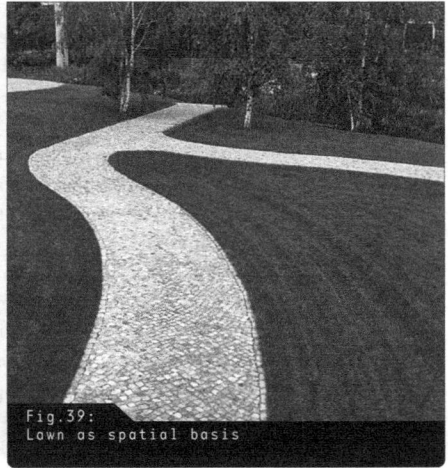

Fig.39:
Lawn as spatial basis

DEVELOPMENT OF LEVELS/HEIGHTS

In open space, differences in height have a strong space-defining potential. A leap in height – the transition between different levels – creates the border to a space. Differences in height can be created by a clear edge or as a gentle transition (modeling). In open areas, spatial levels are often created and structured by plants, through tiering and demarcation between different heights (from lawn to tree). Lawn, meadow, ground cover and herbaceous plants give an impression of flatness and emphasize the horizontal level. Bushes, hedges, larger bushes, solitare woody plants and trees take on a space-defining position as they increase in height. Climbing plants such as ivy and Virginia creeper can create large expanses of green on walls and buildings.

Neatly cut expanses of lawn, expanses of meadow, low ground cover consisting of homogenous herbaceous plants, and woody plants and knee-high displays all have different visual effects as horizontal surfaces due to their different textures, colors and structures. They convey a variety of moods such as solemnity, peace or insouciance. A lawn is the spatial basis for a planned garden. › Fig. 39 With its consistent soft texture, a neatly cut lawn gives a strong carpet-like impression of flatness and imparts visual peace to an open space. The lawn should therefore be as expansive and continuous as possible. Ground features are made clearly visible by the short cut of the lawn. Meadows, on the other hand, change their appearance over the course of a year. The colors of the dominant types of grasses and flowers, changing with the seasons, give the meadow a varied visual

Horizontal
levels

41

dynamic. This is accentuated by the grass waving in the wind. Mowing can be utilized as a means of design, in order to create interesting contrasts, for instance, in the form of lawn paths, mowing of individual areas, or the edges of a meadow. > Chapters Plants as a material, Principles of design, and Plants as a material, Plants: appearance

Ground cover consisting of evergreen, low-growing herbaceous plants and woody plants creates a livelier surface structure than a lawn, due to its varied colors and textures. The smaller the foliage and the lower the plants used are, the more pronounced the effect of flatness is.

> ✎

Space-defining
levels

The elements of space-defining levels are lawns, herbaceous plants, bushes, hedges, large bushes, solitare woody plants, groups of trees and planted facades. In lawn surfaces in gardens and landscapes, bushes and hedges have the function of breaking up and demarcating surfaces, thereby creating spatial depth. > Chapter Spatial structures, Defining space If a visual barrier is desired, bush hedges can be planted with wide intervals between, thereby creating spatial backdrops. > Chapter Spatial structures, Organization of space Bushes also offer a transition from trees to plants at ground level and from open landscape to garden or parkland. The most important characteristic of bushes is their leaves. If they can fill in a large surface area at eye level, their leaves make a considerable contribution to the overall impression made by a display. This may be modest or impressive. > Fig. 40 Bushes remain attractive throughout much of the year (leaf color, blossom, fruits), but often have no pronounced structure.

Aside from trees, clipped hedges are one of the most important design elements, as they introduce a formal component to the design, structure spaces and have a strong contour, structural and textural effect. > Chapters Spatial structures, Borders and Plants as a material, Plants: appearance Clipped hedges can take the following forms:

- Hedged spaces
- Continuous hedges
- Hedge screens

✎

\\ Tip:
Using one plant type, e.g. a ground-covering species throughout the tree and bush display of a housing complex unites different areas and clarifies the plant design.

Fig.40:
Trees and hedges as space-defining
elements

_ Hedge parcels
_ Free-form hedges

Hedged spaces are created by hedge plants that reach above the eye
level of the viewer. Spaces can be created that enclose self-contained plant
themes. Continuous hedges can be grown to half-height, or be arranged
in fan-shapes, curves and other whimsical forms. Hedge screens provide
the open space with green edges or "screen walls." They may stand de-
tached from or in contrast to architecture, but this is not inevitable. Hedge
parcels are height-tiered volumes of cubic greenery. The several rows give
the hedges a greater depth. Hedges clipped into a free form have a strong
sculptural effect. Clipped hedges are a style-defining element in the plans
of the Belgian landscape architect Jacques Wirtz.

Vertical
orientation Vertical surfaces such as building walls and free-standing walls, and
also vertical elements such as pergolas, loggias and screen walls create
spatial boundaries, in closed or transparent form. These can be partially or
entirely planted with climbing woody plants, thereby becoming green and
flowering spatial walls. In planting the whole surface of a building wall

Fig.41:
Planted facades become a spatial contour.

or free-standing wall, climbing woody plants create an interesting tex-tural effect. The green cover appears like a garment. › Fig. 41 Partial planting with climbing woody plants, on the other hand, accentuates certain areas. Pergolas, arcades and trellises are surrounded by plants and create soft transitions and pleasant details. Buildings are given an unmistakable ap-pearance. Climbing woody plants can cover vertical elements while taking up the minimum of space. They are divided according to their mode of climbing. Self-clinging woody plants can overgrow vertical surfaces and elements (and also horizontal surfaces) without aid. Twiners and vines re-quire aids in climbing. › Fig. 42 and Tab. 3

However, a wall-type surface can also be created using hedge plants produced in tree nurseries. In choosing a type of tree, the desired final height, degree of foliage thickness, and whether greenness is required in winter or summer are the decisive factors. Hedges must be cut at least once a year in order to maintain their shape and density. For greater dis-tances, tree walls should be used as boundaries, in order to achieve visual impact.

Growth form	Botanical name	English name	Whole-area cover	Partial cover	Height reached in m	Growth rate*	Ever-green	Decidu-ous
Self-clinging	Hedera helix	Ivy	x	x	10-20	s	x	
	Hydrangea petiolaris	Climbing hydrangea		x	8-12	m		x
	Partheno-cissus quinquefolia "Engelmannii"	Engelmann Virginia creeper	x		15-18	f		x
	Partheno-cissus tricus-pipidatra "Veitchii"	Veitch Japanese creeper	x		15-18	f		x
Twiners (with climbing support)	Clematis montana (all varieties)	Anemone clematis		x	5-8			x
	Clematis montana "Rubens"	Rubens Anemone clematis		x	3-10			x
	Clematis tangutica	Leatherleaf clematis		x	4-6			x
	Clematis vitalba	Traveler's joy		x	10-12			x
	Partheno-cissus quinquefolia	Virginia creeper	x		10-15	f		x
	Vitis coignetiae	Crimson glory vine		x	16-8	f		x
Vines (with climbing support)	Aristolochia macrophylla	Pipe vine		x	8-10	m		x
	Celastrus orbiculatus	Oriental staff vine		x	8-12	f		x
	Lonicera caprifolium	Perfoliate honeysuckle		x	2-5			x
	Lonicera heckrottii	Flame honeysuckle		x	2-4			x
	Lonicera henryi	Evergreen honeysuckle		x	5-7		x	
	Lonicera tellmannia	Honeysuckle		x	4-6			x
	Polygonum aubertii	Type of knotgrass	x	x	8-15	f		x
	Wisteria sinensis	Chinese wisteria		x	6-15	m		x
Ramblers	Jasminum nudiflorum	Winter jasmine		x	2-3			x
	Rosa by varieties	Climbing rose		x	2-3	m		x

* s = slow-growing, m = medium-growing, f = fast-growing

growth form	climbing support	
self-clinging plants	walls trees surfaces (horizontal, at an angle, vertical)	suckers climbing roots wall
twiners	trellis growing trellis steel mesh vine wires stretched horizontally and vertically	
vines	vine wires, stretched pergola loggia	
ramblers	walls trees	wall

Fig.42:
Growth forms of climbing plants and climbing supports

PROPORTION

Proportions and relationships between values affect a space's ap-pearance. "Proportion" describes the precise (and mathematically cal-culated) relationship between the significant measurement values of objects (e.g. the relationship between height and breadth), and is de-scriptive of the visual weighting of design components' value relation-ships. A space can be shortened or deepened by changing the proportions, but this can also be achieved by means of perspective. In the case of a street, if the proportion of the average overall height of the facades of the buildings to the height of the trees in the street is 3:5, the propor-tions appear balanced and harmonious. Changing the lateral length to create a proportion of 2:6 or 4:4 produces entirely different spatial ef-fects. In providing a proportion reference, trees mediate between build-ings and human beings. It is therefore important that trees harmonize with the height and size of the buildings. High trees require a street with a broad cross-section, and need to be far enough away from the build-ings, while smaller trees look better in narrow streets or closer to build-ings. › Fig. 43 Tricks of perspective can be used to emphasize the depth and breadth of a space if the proportions are tapered. An effective depth can be achieved using interpolated, tiered or linear elements that visually

where tree size remains the same

where tree size changes

Fig.43:
Size relationships between trees and buildings

structure spatial depth. These elements can also influence the proportions of the entire space through particular size or characteristics. Single trees, pergolas, reflective expanses of water etc. are suitable for this purpose. > Chapter Spatial structures, Organization of space Linear structures such as formally clipped woody plants in h edge form or colorful effects such as leaves and decorative blossom can also be used to create perspective.

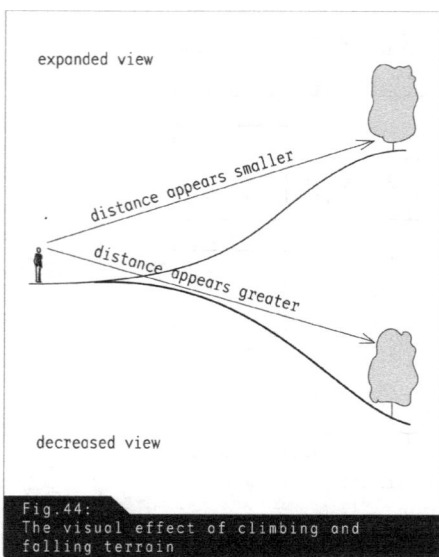

Fig. 44:
The visual effect of climbing and falling terrain

Topography and modeling of the ground can have a significant influence on proportions in outdoor spaces. For instance, an expanse that inclines away from the observer's viewpoint appears longer, because it is canted away from him or her. Falling ground appears broader. Conversely, a climbing expanse appears shorter, as all of it can be seen, and, visually speaking, it is moving toward the viewer. › Fig. 44

\\ Tip:
The colors red and orange bring an object into the foreground and therefore have a visually shortening effect. Blue, blue-green and violet blue displace an object into the distance, and therefore have a visually distancing effect (see Chapter Plants as a material, Plants: appearance).

\\ Tip:
Park layouts, town squares and gardens offer numerous opportunities for studying plants and their use. Observing and analyzing plants makes their diversity and their different qualities evident. What character does a plant have? How does it harmonize with its surroundings? By analyzing good and bad (!) examples, one can work on one's own planting plans in a more informed way.

48

PLANTS AS A MATERIAL

Choosing appropriate plant species is a process in which a number of criteria must be weighed up. It is important to combine the various appearances of the plants with each other and with their surroundings harmoniously, thereby creating a clear image with a strong impact. It is important not to lose sight of the overall effect of the open space layout. Well-founded knowledge of plants is a requirement for plant-based design. For instance, strongly growing plants can rapidly out-compete their neighbors and completely overgrow areas of a garden. Knotgrass (Polygonum), for instance, has this property. Other plant types grow very slowly, or are intolerant of neighboring plants that reach a similar size.

Plants are a living material and often develop in an unplanned way. Our planning presents a basic framework, within which the selected plants will develop. With regular care, the development and quality of an open space can be directed.

PLANTS: APPEARANCE

Effective designs are often very simple and consist of a few carefully chosen plant species and varieties. A precise conception of the appearance of the plants used and the application of aesthetic rules are required. Knowledge of these principles is a more reliable guide in the selection and positioning of plants than simple intuition.

Form

Forms are created by the planted sections that border them. Depending on the density of the outermost tips of growth (twigs, leaves, blossoms), the form or the structure of a plant will be more visible. In the summer, deciduous trees and bushes are as noticeable as evergreen trees and woody plants, especially if they have thick leaf cover. Some deciduous trees and bushes have a definite shape in winter, if the branch and twig system is closely arranged and creates striking contours. The more simple and unambiguous the styling of a plant is, the easier it is to comprehend, to describe, to draw, and to designate. The classification of plant forms distinguishes between different types of growth. › Fig. 45 Forms may consist of body or surface. Many continuous forms derive from the simplest basic forms: square, circle and triangle. Free forms, on the other hand, are significantly more complicated. Their character can be seen, for instance, in the separated, thickly needled ends of a mature cedar's branch system.

Formal characters

Aptly formed woody plants stand out due to their architectonic or graphic effect, and can structure a park, garden or display. Through the form and direction of their growth, they create either a static or dynamic impression. Formal characters can be divided into "without direction,"

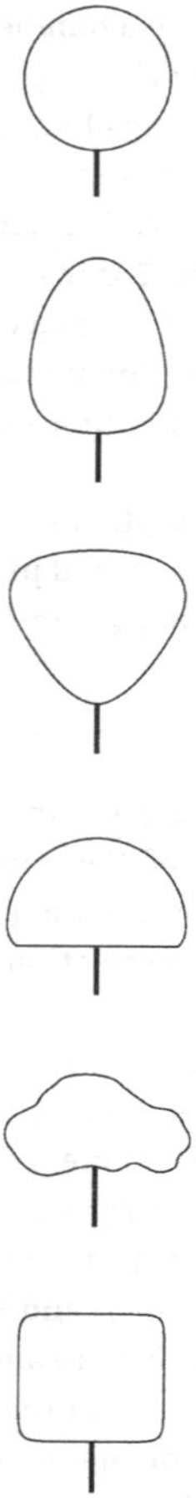

	Type	Example	Use
	Ball-shaped	Acer platanoides "Globosum" (Norway maple)	Compact small trees for enclosed spaces, front gardens
	Egg-shaped	Tilia cordata "Erecta" (small-leaved lime)	For formal situations with tree rows and avenues, urban open spaces
	Funnel-shaped	Prunus serrulata "Kanzan" (Japanese cherry)	For tree rows and grids
	Umbrella-shaped	Catalpa bignonioides (Indian bean tree)	Fully grown trees for sheltered seating or small areas requiring shade
	Pine-shaped	Populus nigra "Austriaca" (formally clipped Austrian black poplar)	Silhouette with a powerful effect in open landscapes with hills and mountains
	Box-shaped	Tilia platyphyllos (formally clipped broad-leaved lime)	For formal situations, green architecture

Fig. 45:
Tree form types

"with a fixed direction," and "with non-constant direction." The sphere as a simple design shape is without direction and has a static effect. Horizontal and vertical plant forms are static with a fixed direction, whereas a climbing or overhanging plant has a non-constant direction. It radiates

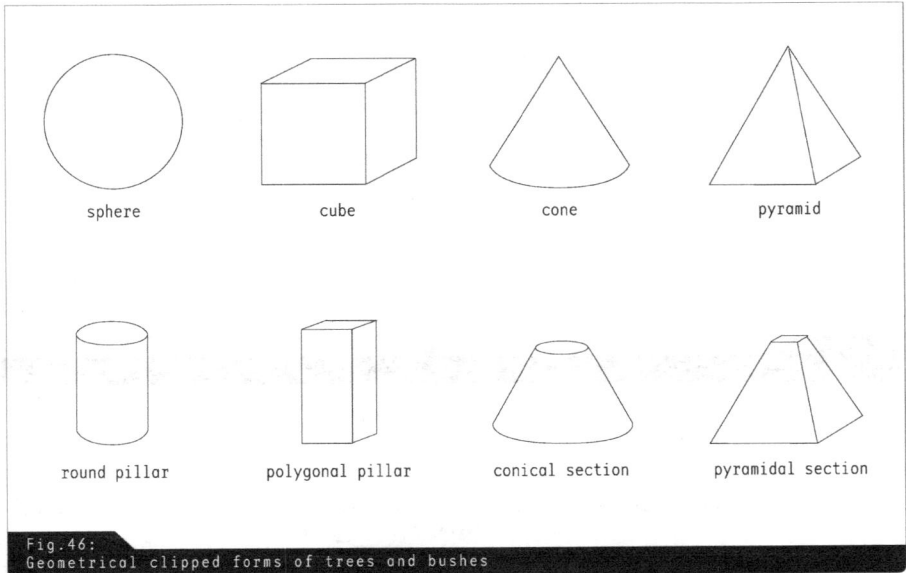

sphere   cube   cone   pyramid

round pillar   polygonal pillar   conical section   pyramidal section

Fig.46:
Geometrical clipped forms of trees and bushes

movement, creating a visual dynamic. A static or dynamic impression is strengthened by different and contrasting combinations of formal characters. For instance, a vertical form (a pillar) standing by a curved path (non-constant direction) appears as a contrasting fixed form. Horizontal plant forms (tree rows) create a recumbent counterpoint to rising buildings (high-rises), and sphere-shaped plants (without direction) may flank a contrasting curved ribbon of plants (non-constant direction). › Chapter Plants as a material, Principles of design

Formally clipped woody plants

Regular formal clipping gives trees, bushes and hedges continuous clear outlines. Selected deciduous and coniferous species are cut into geometrical (cube, pillar, sphere, pyramid, cone, conic section etc.) or organic figures. › Fig. 46 Strictly clipped trees are divided according to whether they are box, roof, trellis, or ball-shaped. › Fig. 47 Formally clipped hedges create continuous, clear spatial edges. Low formally clipped hedges structure garden areas without closing them off visually. Formal clipping controls the volume of the plants and keeps it nearly constant. Formally clipped woody plants are particularly good for garden layouts with an architectural character and the structuring of open layouts. › Fig. 48 However, formal clipping is only possible for a limited number of plant species. › Tab. 4

51

spherical shape

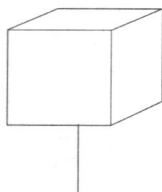
box shape

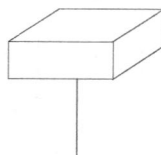
roof shape

espalier shape

Fig.47:
Clipped forms for trees

Tab.4:
Trees and bushes suitable for formal clipping

Botanical name	English name	Solitaire tree	Hedge	Arch	Geometrical body	Umbrella form	Trellis	Bonsai
Deciduous trees and bushes								
Carpinus betulus (and varieties)	European hornbeam	x	x	x	x	x	x	x
Cornus mas	European cornel		x			x	x	x
Crataegus (species)	Hawthorn	x						
Fagus sylvatica	European beech		x	x	x			
Platanus acerifolia	Hybrid plane	x						
Tilia (species)	Lime or linden	x	x	x				
Evergreen trees and bushes								
Buxus sempervirens arborescens	Tree boxwood		x		x			
Ilex aquifolium (in varieties)	Holly				x			
Ilex crenata (in varieties)	Japanese holly				x			x
Ligustrum vulgare "Atrovirens"	Wild privet		x		x			
Pinus (species)	Pine					x		x
Taxus (species)	Yew	x	x		x			x
Fruit-bearing woody plants								
Malus domestica (in varieties)	Apple						x	
Pylus communis (in varieties)	Pear						x	

Fig.48:
Formally clipped bushes

Fig.49:
The branching character of deciduous
trees is visible in winter.

It is important to know how visually effective the form of a plant is
from a variety of distances, as in landscape architecture, the distances
from which the displays can be viewed are defined. The visual effect of
the form changes with distance. From further away, the eye registers a
silhouette-like impression rather than a form. From a medium distance,
the vegetation appears to have more body due to the effect of shadows.
Viewed close to, the color and texture of a plant have a greater visual
effect than its form. Distances should also be taken into account when
deciding on the number of different species. At a long distance, the casual
observer can only register a few different species in a large group of
trees.

Like its form, the plant's characteristics contribute significantly to
its appearance. They represent its characteristic growth type. In landscape
architecture, trees are among the plants whose characteristics are most
noticeable. In trees and bushes, they can be most clearly seen in winter.
› Fig. 49 In classifying plants according to characteristic types, they are
shown diagrammatically, as if clipped. › Fig. 50 Division into types by form
and characteristics creates an overview of which plants will suit a cor-
responding situation visually. A tree with regular, solid-crowned growth
is suited to a formal context in which trees are placed at regular intervals
(a town square). A tree with loose, irregular growth can lighten a hard and
uniform building facade.

53

	Type	Example	Use
	Round, spherical	Platanus acerifolia (mature plane)	Formal situations with rows, avenues and grids
	Round/ egg-shaped	Acer platanoides "Cleveland" (Norway maple)	Open urban locations, including squares, streets and park layouts
	Irregular, loose-crowned	Gleditsia triacanthos (honey locust)	Informal situations as a single tree, in mixed displays
	Multi-trunked	Acer palmatum (Japanese maple)	In connection with buildings, for emphasis
	Cone-shaped	Corylus colurna (Turkish hazel)	Group plantings or as a focal point between other plants
	Pillar-shaped	Populus nigra "Italica" (black (Lombardy) poplar)	Open landscapes, flat expanses and gentle rises, to emphasize linear elements (avenues). A contrast to markedly horizontal constructed elements and entrance areas
	Overhanging	Betula pendula (silver birch)	Solitaire tree with artistic form for solo placements and loose groups, for scenic park layouts and buildings with elaborate diverse forms

Fig.50:
Different characteristics of trees

✎
\\Tip:
Studies using models allow different combinations of characteristics, branch structures,
textures and form types of trees and bushes to
be tested. Plant materials such as twigs, dried
flower and seed sprays and fruits are useful
for this.

Herbaceous plants and grasses may also show distinct characteristics. Blossoms, leaves, stalks and the growth direction of shoots create different growth forms: herbaceous plants with a single shoot have
leaves low to the ground and a single flowering stem, e.g. mullein (Verbas-
cum) or foxglove (Digitalis), while straight-growing clump plants grow
stiffly upwards, e.g. iris and Chinese silver grass (Miscanthus). Inclining
clump plants diverge in soft arching lines, e.g. day lily (Hemerocallis) and
fountain grass (Penisetum). The individual growth forms have different
effects; stiff, upright herbaceous plants give an impression of structure
and accentuate, while inclining herbaceous plants appear gentle and el-
egant. When a variety of plants are planted together, their ability to combine in design terms can be considered with regard to the characteristic
growth pattern and form of each one.**>** Chapter **Plants as a material, Principles of
design**

The development of a plant's characteristics is strongly dependent
on light and competition. A plant that prefers a sunny habitat would not
develop its normal characteristics in a shady place; the result would be
stunted growth and a failure to flower.

Texture

Texture is one of the most formally effective properties of a plant.
Both the density of the whole plant and the surface qualities of the individual leaf, the stems and shoots create a textured effect. What texture
means is the characteristics of the plant's leafage: the form and surface
qualities of the individual leaves, their size, their alignment, their number,
and the way light reflects from their surface. The delicacy of twigs and
shoots also gives a plant texture. A simplified system of textures ranges
from "very fine"(lawn), "fine," "medium," "coarse," to "very coarse." **>** Fig. 51
A cut hedge, like a closely mown lawn, generally has a dense, fine and
"smooth" surface with a harmonious surface and wall effect. For instance,
the dense, fine texture of a cut yew hedge and its architectonic form give
a formal, strict impression, while a free-growing rosebush emphasizes the
natural. If plants appear in connection with buildings or other structures,
the existing or planned material textures and structures of the building
should be taken into account when utilizing plant textures. For instance, if

Fig.51:
Examples of different textures: fine, medium fine, medium coarse, coarse

Fig.52:
Structure created by planting grosses in a grid

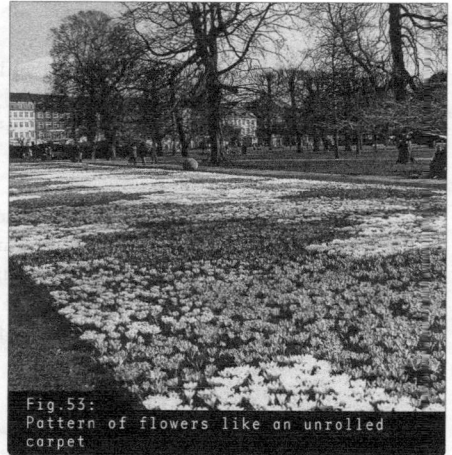
Fig.53:
Pattern of flowers like an unrolled carpet

the leaves of a plant are of the same size as the bricks in a wall, it is easy for the visual impression to be dull. › Chapter Plants as a material, Principles of design Plant textures have a variety of abilities:

_ They can lend the vegetation profile an impression of strength and coherence.
_ They can create accentuated effects.
_ Fine textures can create a harmonious and clear background, visually enlarging the garden space.
_ They can serve as support in order to emphasize the depths of the landscape.
_ They can create an impression of unity in a planting, if the same texture is continued through a line of plants of different species.

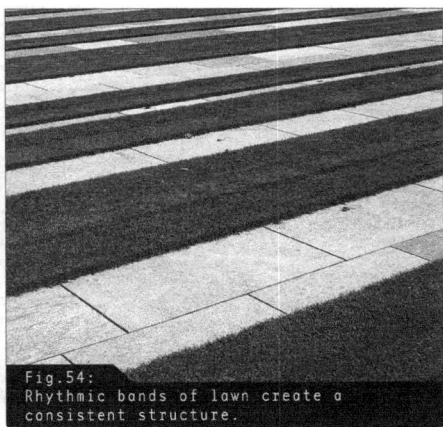

Fig.54:
Rhythmic bands of lawn create a
consistent structure.

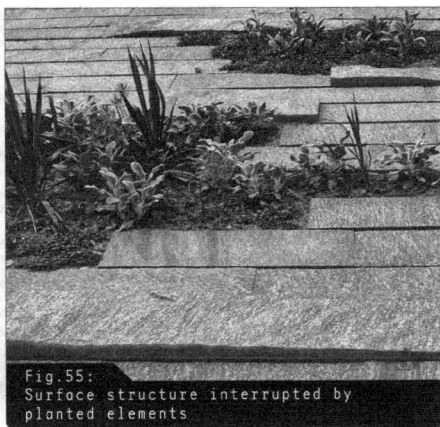

Fig.55:
Surface structure interrupted by
planted elements

Structure

Structure is defined as the interior construction of a design unit. Structure arises through the repetition of internal elements. The expression "structure" is applicable to design at all levels. A plant display, a draft or a text requires such a structure in order to be comprehensible.

Area structures

The arrangement of forms of the same or of similar kinds in large numbers on a surface produces a structural effect. > Fig. 52 Regularly arranged structures are ornamental, patterned (wallpaper, printed fabric, carpets), and emphasize the flat effect. > Fig. 53 Irregularly arranged structures appear livelier and more spatial. > Figs 54 and 55 The textures of materials and plants create very different effects in a flat situation. For instance, bedding areas consisting of only one type of herbaceous plant make a strong impression. > Figs 56 and 57 Bedding areas with ground-covering displays of different summer flowers and herbaceous plant types are further examples of an area structure. A structure within a display can be created, for instance, by plants with pronounced foliage (grasses, ferns), used repeatedly. > Chapter Plants as a material, Principles of design

Spatial structures

Spatial structures are either transparent or interrupted. In a forest with high trees, walkers finds themselves in a structured space. Trees are in front of, behind and on either side of them, branches and twigs are above. It is a multitude of identical and similar elements, and their division that results in a spatial structure. > Figs 58 and 59 In order to give a garden a spatial structure, a "framework," identical or similar space-defining elements (e.g. trees or woody plant forms) are selected and integrated into the space recurrently. These arrangements may be dense, airy, even, rhythmical or unordered, and may have a variety of effects. > Figs 60, 61 and Chapter Plants as a

57

Fig.56:
Surface structure created by uniform
carpet of flowers

Fig.57:
Surface structure created by regular,
linear arrangement of cushion-like bushes

Fig.58:
Spatial structure created by row of
vertical plant bodies

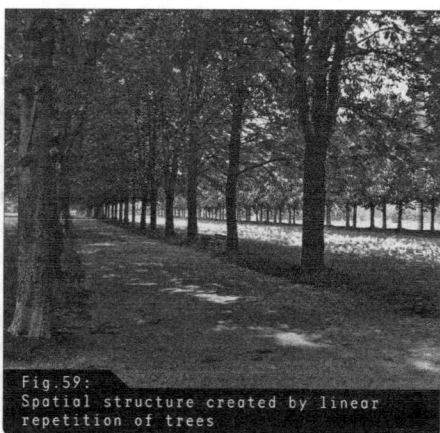

Fig.59:
Spatial structure created by linear
repetition of trees

material, Principles of design In deciduous woody plants, while bare of leaves, and some coniferous woody plants, the spatial framework, the branching characteristics, are visible. The resulting linear, graphic impression given by the branches can be used to good effect in creating background contrast.

> Chapter Plants as a material, Principles of design

Contour   The outline or the silhouette of a plant is described as its contour. A distinction is made between woody plants with a continuous or an open contour. Due to their regular cut, formally clipped woody plants and

58

Fig.60:
Fluid interplay of plants

Fig.61:
Different plant typologies structure a
space horizontally and vertically.

formally clipped hedges have a dense texture, and clear, continuous contours adhering to a line. These are important for structuring a garden. › Fig. 62 Formal garden layouts are unthinkable without formally clipped woody plants. Thickly textured free-growing woody plants and woody plants with an organic formal cut are also continuous and clearly delineated. The plastic and visual impression they give is "heavy." › Fig. 63 Open contours are either arranged, e.g. by layering of branches (table dogwood, or Cornus controversa) and uniform tiers (Serbian spruce, or Picea ormorika), or irregular and loose. The closer the observer's viewpoint, the more clearly he or she will perceive the contours of the individual leaves.

Color

In a garden, most observers notice the colors and textures of flowers, leaves and fruit, although characteristics and form are the most important visual factors in a plant's appearance. The slow maturing of a garden shows in the characteristics and form of plants, while the color and texture of different species emphasize seasonal change. Plants possess a great diversity of color, which is added to by the variations resulting from lighting and texture of leaf and blossom surfaces (e.g. lustrous, matte etc.). Colors can be systematized, distinguished according to their effects and tested using color wheels and color tables. Color tone, lightness of color and luminosity determine the impression created by a color. When working with colors in landscape architecture, it should be borne in mind that in most gardens and landscapes, green is dominant, changing to brown in autumn and winter, while the other colors cover only a very small area.

Lightness of
color

Color is dependent on light. Type of light, light strength and angle of incidence are crucial to its effect. The coloration of plants appears entirely

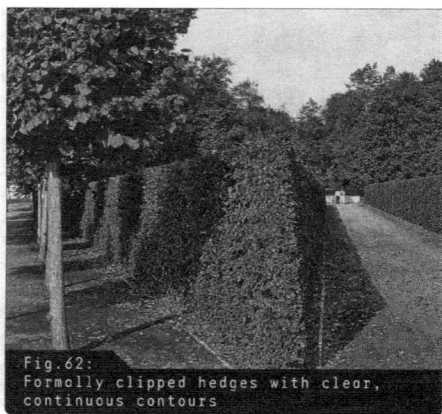

Fig.62:
Formally clipped hedges with clear, continuous contours

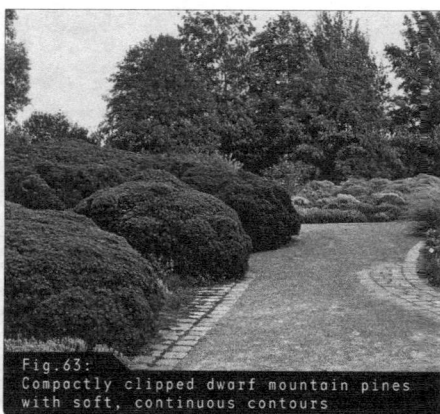

Fig.63:
Compactly clipped dwarf mountain pines with soft, continuous contours

different in sun and shade. During planning, it should be determined which areas of the open space will receive sunlight during which parts of the day. Diffuse light reduces color intensity, which increases with direct light exposure. A clear or cloudy sky will therefore affect the appearance of blossom and leaves, as will artificial light. In daylight, yellow and yellow-green have the strongest intensity. At night, blue-green does. In bright light and in increasing darkness, colors grow paler, although the quality of lightness remains visible. The impression of depth in a display also changes with the direction from which the light is coming. Morning and evening sunlight (light coming in from the side) creates a much stronger effect of spatial depth in a garden or landscape than at midday. Diffuse light also creates less depth.

Every color has its own specific lightness. Blue is dark, yellow is light. Red is moderately light and somewhat darker than orange. Color tones can be altered by the addition of black or white. The graduations produced can be represented in a color wheel. › Fig. 64 The intense spectral colors are represented in the outer ring of the color wheel, while the inner ring displays a surface that represents the maximum lightness as "neutral" white, or maximum darkness as "neutral" black. Between the outer circle and the neutral center, a graduation of color intensities takes place. If blue is darkened, it becomes "heavy" and loses its ethereal character. If yellow is whitened, it loses its brilliance and becomes pale. The more a blossom color is muted by the introduction of gray, the less brilliance and long-distance effect it has. Optically, pure colors therefore appear nearer than broken ones.

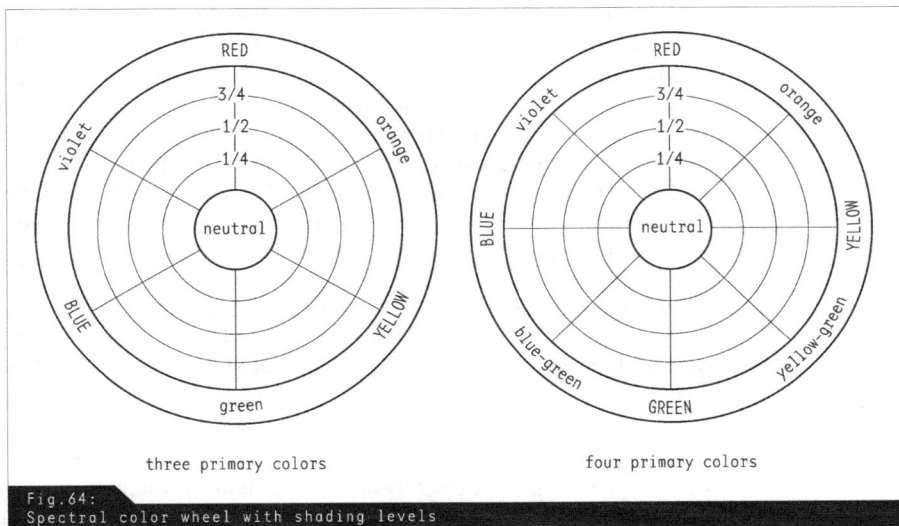

Fig. 64:
Spectral color wheel with shading levels

three primary colors   four primary colors

White strengthens the effect of all colors. Displays with white flowering plants and variegated (white-edged, white-flecked) foliage can lighten shaded areas. Silver-gray plants have a similar lightening effect, especially when combined with white ones.

Complementary colors

Complementary colors are opposite each other in the color wheel. The secondary colors of green, violet and orange are created by mixing the three primary colors, red, yellow and blue. These six colors are the spectrum formed by breaking up white sunlight. A primary color is always opposite a secondary color, e.g. red is opposite green. In spectral mixes, opposing colors create white. Complementary colors heighten each other's effect, their color intensity. Red is brilliant against a green background, yellow against a violet background, blue against orange and vice versa. Each color has the tendency to push other colors towards the pole most opposed to it; green makes yellow look redder – that is, green creates its opposing color of red in yellow. Yellow mixes with the non-existent red created by complementary contrast and moves toward orange; blue makes green look more yellow; green makes blue appear violet; and yellow creates a bluer-looking green.

For the four primary colors of red, yellow, blue and green, the opposing colors, which create the strongest effect, fall exactly on the opposing place in the color wheel. › Fig. 64

The color wheel can be divided into warm and cold colors. Red, cr-
ange and green are associated with warmth, as are yellow-green and bright
leaf-green. Blue, blue-violet and blue-green are considered cold colors.
Warm colors appear to be closer to the viewer, whereas cold colors recede
into the background visually, causing open space to appear deeper than it
is. Mid-green and blue-purple are considered neutral colors. The green of a
landscape therefore has a calming and stabilizing effect. A color combination consisting only of warm or cold colors appears harmonious, while a
combination of warm and cold colors creates contrast, without necessarily
being discordant.

Where a harmony exists, one can take a color tone and vary it using
gradations of closely related tones, for instance in a warmer or colder direction. Background, surroundings, growth form and texture of neighboring plants as well as the color variation itself determine its quality. The
numerous degrees of green in woody plants, knowingly juxtaposed, may
be all that is required for good color harmony, and to bring peace into
a display. Combining dark, coniferous woody plants with lighter, deciduous woody plants intensifies color tones. In traditional Japanese gardens,
a uniform color harmony is created using predominantly evergreen trees
and bushes. Further colors only appear in the spring with the coming of
the cherry blossom and in autumn with the autumnal leaf colors, and are
effective precisely because of their brief appearance.

Two-part harmonies are colors that are opposite one another in the
color wheel (complementary colors): orange and blue, golden-yellow and
ultramarine, red-orange and blue-green. › Fig. 65 Three-part harmonies are
colors that are a third of the color wheel distant from one another: blue,
yellow and red, or ultramarine, red-orange and yellow-green. One further
possibility is to incorporate a neighboring color of one of the opposing
colors into a two-part harmony (ultramarine, yellow-orange, yellow), or to

\\Example:
Plant beds with only one blossom color are
described as monochrome color compositions. In
single-color plant beds, contrasts of form are
particularly effective, e.g. in flower sprays:
yarrow (Achillea), goldenrod (Solidago) and
black-eyed Susan (Rudbeckia) standing close
together in large groups.

\\Tip:
Even the use of a single color permits a wealth
of color tones created by lightening, darkening,
making the color warmer or colder. Restricting
the initial range to a few tones creates an inner
coherence and prevents brash, random colorfulness. Powerful colors work to best effect in sunny
areas, while subdued, light colors are most effective in shady areas.

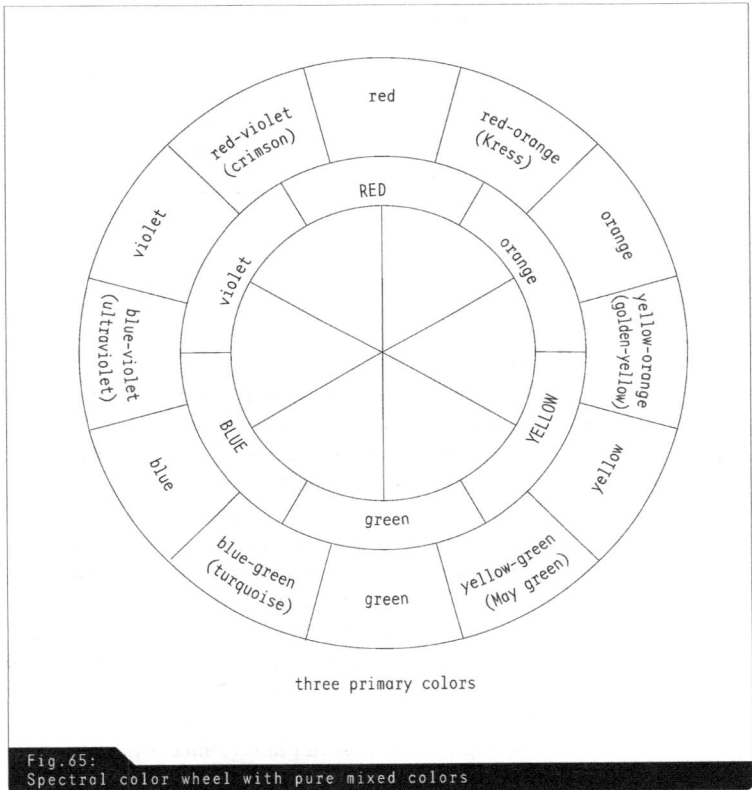

three primary colors

Fig. 65:
Spectral color wheel with pure mixed colors

leave one color out of two complementary pairs (blue-green, red-orange, orange). When two primary colors are combined with a secondary color, they create a strong effect (red, blue, violet). Two secondary colors with a primary color have a more subtle effect (green, orange, red). There is, however, a wide range of color tones existing within these conceptual color harmonies – every red does not harmonize with every green. If two color tones do not harmonize, a third tone may relieve the dissonance. With every tone added to a three-part harmony, it becomes more difficult to create an expressive image.

Effective color harmonies can be created in combination with silver-gray foliage. Strong colors such as red and blue become even more brilliant in this environment, while restrained colors and pastel tones also show to best effect. › Tab. 5

Two-part harmony	Three-part harmony
blue – orange	blue – red – yellow
yellow-orange – ultramarine	blue – red – silver-gray
orange – silver-gray	light blue – yellow – silver-gray
rose – silver-gray	yellow – white – silver-gray

TIME DYNAMICS

Planning to use plants as a design element goes beyond flat and three-dimensional space to include the fourth dimension: time. In contrast to concrete and stone, plants are a living material whose form changes due to growth. The tempo of this change can vary greatly. To some extent it can be viewed from day to day, especially when leaves, flowers and fruits are developing. Depending on lifespan, a plant growing in our latitudes changes rhythmically through the seasons and over the years, decades or centuries. In a garden, a continuous coming into being and passing away can be observed. However, this innate dynamic also poses questions. When is a garden complete? When is it gaining, and when does it start losing quality? Planning to use plants as a material also means being prepared for a long timescale. Growth requires time. A new open-space layout with plant elements has a sparse, incomplete appearance compared to a garden that has been growing in place for years. It can be disappointing for planner and user, if they fail to take into account the element of time. When executing such a layout, it is therefore desirable to choose plant specimens large enough for the space's proportions, in order to create space and structure at an early stage. The same principle applies: open spaces are structured by the characteristics and form of plants, while the color and texture of the different species emphasize seasonal change.

Seasonal change   While the spatial structure of many woody plants is stable, the changing color profile early in the year and in autumn often has a dramatic effect. › Fig. 66 Every plant species has its own typical series of seasonal changes. The mass of evergreens and deciduous woody plants changes over the seasons. In summer, deciduous woody plants form the main framework of a display, while in winter the evergreen and coniferous woody plants come to the foreground visually and may form a framework for the display if planned correctly. They change in appearance the least, creating stability. A rhododendron bursts into glorious color in May, is less noticeable in the

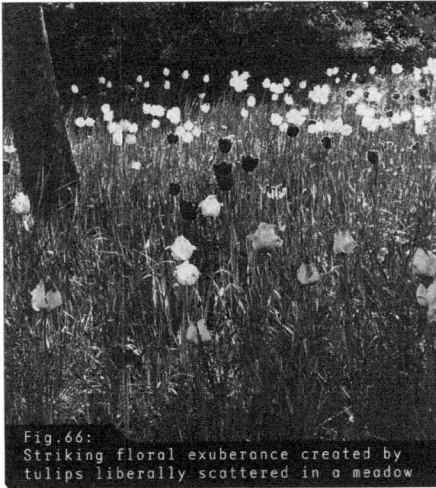

Fig.66:
Striking floral exuberance created by
tulips liberally scattered in a meadow

summer and becomes visible again in winter due to its evergreen leaves. After their leaves have fallen, summer-green woody plants become linear and graphic. › Chapter Plants as a material, Plants: appearance Herbaceous displays change their appearance in an especially marked way. In winter, the above-ground parts of many herbaceous plants die off, reappearing in spring in the same place and then increasing considerably in height and volume. Choosing plants for their appearance throughout the seasons is recommended, in particular for any garden layout that will be viewed throughout the year, for instance a house's garden.

When developing a plant schema and arranging the plants, it is important to allow for a continuous series of colors, from early spring to late autumn. One option is to divide the plant groups according to their time of flowering and arrange them in different parts of the plot or landscape, as a multiplicity of flowers at the same time in the same place tends to weaken the overall impression and make the layout look badly thought-out. In a natural landscape, colorful periods are usually short episodes, followed by quiet periods. The effect of "eruptions of color" can be strengthened by choosing plants whose colors have a connection with the seasons. To achieve this, one would choose plants with yellow and blue flowers for the spring, deep blue, white and pink flowers for the early summer and scarlet, dark red, violet and deep yellow flowers for the late summer. Brown and violet leaves and flowers are suitable for autumn, deep green and brown leaves and red berries for winter.

One important point to consider in planning a display for seasonal effect is that many plants with effectively colorful blossoms do not contribute to the structuring of a garden area. The lilac, for instance, is a plant with beautiful flowers, but unremarkable leaves and branch structure. A stronger visual frame may balance out the effect of plants with nondescript foliage and branch structure. Taking the lilac as an example, this would involve putting a low, dense or clipped hedge in front or using the lilac as a background for other plants that make their seasonal contribution at other times, when the lilac is not blooming. Hybrid tea-roses are another example of a plant with fine flowers, but a foliage and branch structure that is not very decorative, especially when clipped. Rose gardens are therefore often formally laid out in such a way as to bring out the special nature of the flowers, while visually toning down the bare stems and foliage through framing, generally with evergreen, low, clipped hedges. › Fig. 67 When in bloom, the different colors of the rose flowers are made more distinct by the framing hedges. Beds of roses may also be enhanced by herbaceous plants planted in between. These, however, must appear as companions to the rose flowers rather than competing with them visually.

Many trees of small stature with striking blossoms, such as ornamental cherry, only have a short flowering period. They should be considered when planning a small-format garden layout, or limited or closed garden spaces such as an atrium courtyard. Mixed with other trees and bushes of different heights, they can appear with a brief burst of color in different

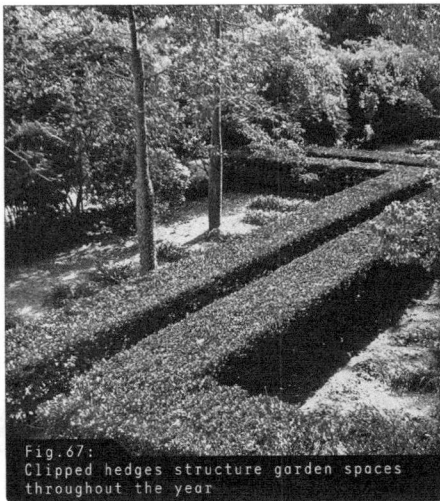

Fig.67:
Clipped hedges structure garden spaces throughout the year

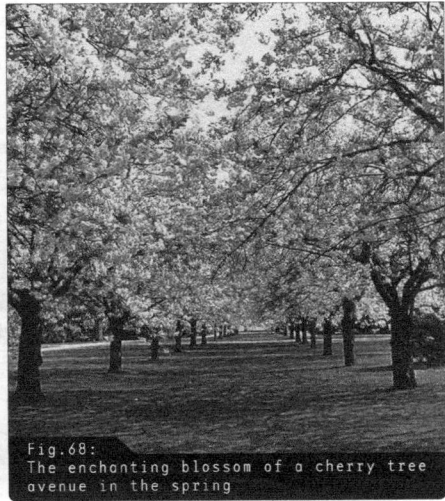

Fig.68:
The enchanting blossom of a cherry tree avenue in the spring

places at different times. A further, effective possibility for revealing seasonal changes is to plant a number of flowering trees as standards in rows, avenues or grids. › Fig. 68

Bedding plant
displays

The main purpose of bedding plant displays is to achieve strong color effects. Annual, non-hardy summer flowers must be grown or acquired anew every year. Preparation, planting and maintenance (continuous provision of water and nutrients) are labor- and cost-intensive, but are justified in the case of carefully chosen, imposing and much-frequented places such as the areas in front of imposing buildings, public spaces, pedestrian areas, historical gardens and parks, civic parks and special gardens. › Fig. 69 In simple villages, in farm gardens or as ornamental plants in tubs or boxes, annuals usually play the leading role. Well-designed and tended bedding plant displays can contribute significantly to the positive image of a town.

Bedding plant displays of spring, summer and autumn flora create a seasonal diversity. There are also places within the landscape where their use may be appropriate, if the design is sensitive and harmonizes with the local or regional landscape. This is achieved by choosing plants whose color and arrangement is based on the natural ground-cover vegetation mosaic (e.g. forest or heather). Late summer is a fairly quiet period for the appearance of parks and landscapes, and can be made more interesting by the planting of annuals that bloom during this time.

Fig.69:
Carpet of bedding plants at a regional garden show

Autumn colors, created by the leaves of trees and bushes, are a welcome change before the arrival of winter. To achieve the greatest visual effect, plants with a seasonal effect may be combined with evergreens and coniferous plants, or those that shed their leaves late in the year. This allows the warm autumn colors to be emphasized by the colder green ones. The warm, strong yellow, orange, red and crimson colors of the leaves remaining on the trees and woody plants, and lying on the lawns and paths bring about an almost impressionistic effect in a garden or park. Beneath single trees and solitaire woody plants, a large area of wonderful color may be created for a short time, if these plants are allowed to stand freely. Plant species with conspicuous berries are less effective in late autumn and early winter, unless they have evergreen plants as a background. In order to achieve the necessary visual effect, late-shedding plants and evergreen species planted for their berries should be acquired in great numbers. In smaller spaces, such as a garden, a single plant may achieve this effect.

In winter, the textures of plants are emphasized primarily by frost and snow, especially filigree plant forms such as grasses, ferns, and the fruiting sprays of herbaceous plants. > Fig. 70 These should therefore only be cut back at the end of winter. All in all, winter offers a small range of visually interesting plants compared with the other three seasons. In some species, the color of the branches is a valuable addition to the effect created by evergreen vegetation and colorful berries, which do not remain for the whole winter. For instance, the striking red twigs of the

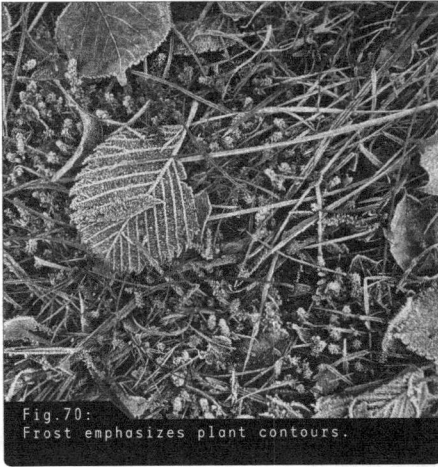

Fig.70:
Frost emphasizes plant contours.

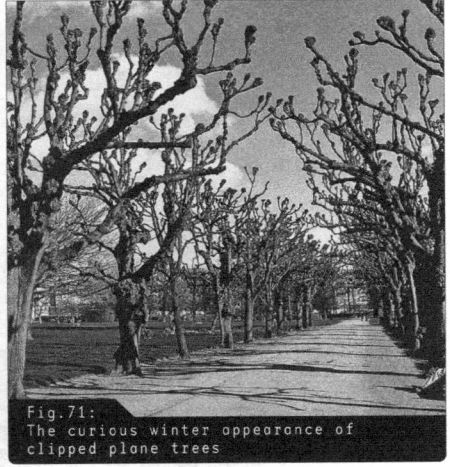

Fig.71:
The curious winter appearance of clipped plane trees

dogwood type Cornus alba "Siberica" may make a winter landscape distinctive. This effect is strengthened in combination with white-trunked birches and evergreen woody plants. The red twigs can be very effective when the window frames of the building are of the same color. The yellow twigs of the green-barked dogwood (Cornus stolonifera "Flaviramea") have a similar effect. The branch and twig systems of some trees and, to a lesser extent, some bushes, create a visually interesting effect in winter. › Fig. 71 Plants should be chosen in such a way that these species have the sky as a background when seen from a variety of angles, or else have a simple background, for instance a wall or house wall, in order to achieve a good graphic effect. › Chapter Plants as a material, Plants: appearance Another visual effect results from the twigs of mature trees in a grid planting just touching each other against the sky. Taking winter into account during planning is one factor that is often neglected.

Development in size and life cycle The appearance of plants is subject to changes throughout their lifespan. The speed of their growth, and therefore the changes in their form, are determined by which of the following groups of organisms they belong to:

_ Trees
_ Bushes
_ Herbaceous plants
_ Bulbs
_ Annuals and biannual (summer flowers)

Fig.72:
Adding young trees to a path demonstrates the growth development and lifespan of trees.

Trees

Trees have a long lifespan and grow relatively slowly. › Fig. 72 They provide continuity across the seasons and through the decades. In gardens and landscapes, they are the most significant and "abiding" element. They provide spatial and temporal continuity and presence. Trees connect a town with the surrounding landscape, and connect a town's districts and buildings. In urban open space in particular, slow growth must be taken into account in plant design projects. The excessive shading of herbaceous plants, bushes and lawns beneath tree crowns causes the roots of the plants to compete for water and nutrients and the shaded plants to grow sick or die.

Bushes

Bushes also increase in size slowly over the years, but do not attain the same age and size as trees. They serve to structure the ground surface and create demarcations. › Chapter Spatial structures, Borders Bushes create a visual connection between trees and ground plants; a park with only trees, herbaceous plants and lawns would appear very open and have little spatial depth. Bushes create a transition between a park or garden and the open landscape.

Herbaceous plants

Herbaceous plants are perennial plants whose aboveground parts, unlike those of trees and bushes, die off following the autumn. In the spring, herbaceous plants re-emerge from frost-hardy underground storage organs. The height reached by herbaceous plants ranges from carpet-like shortness to over 2 m. The seasonal change of form dynamizes a garden or open space considerably.

Bulb plants

Geophytes – plants with underground storage organs such as bulbs, tubers and roots – spend most of the year invisible, buried in the soil. Many appear early in the year, when the leaves of trees and bushes have not yet sprouted and the sun still reaches the ground. Their natural habitat is forest or forest edge. Their leaf is short-lived and withdraws after flowering. The withdrawal of the leaf is essential to the health and flowering ability of the plant. In order to conceal the leaf, they are better sited in flat plantings than in lawn surfaces. They should also be long-lived and compete with other plants neither visually nor ecologically. Late-blooming bulb plants (e.g. tulips) come from dry areas in which there is only sparse vegetation and therefore little competition. For this reason, they do not combine well with herbaceous plants and bushes. Examples of geophytes that are suitable for lawn surfaces are crocuses and narcissi. Their leaves should be removed at the earliest after they have turned yellow.

Annuals and biennials

Annuals last a single growing season. They are normally used for temporary displays. If they do not show competitive behavior, they may be combined with herbaceous plants.

Biennials last for two growing seasons. They usually flower in the second year and produce many seeds before dying off.

PRINCIPLES OF DESIGN

In order to create a good planting plan, the different appearances of plants – size, form, color and texture – must be combined to form an inner coherence. This requires a uniting idea, a main theme. Theme ideas form the content of the design, which is given shape by space, plants and materials. Knowledge of universally applicable principles of design, such as contrast and balance, repetition, rhythm and order etc. gives us the tools to make our ideas clear and recognizable, independent of images of exemplary plant combinations from numerous gardening books.

Contrasts

Contrast is one of the most important principles in design using plants. It is required to create the tension and attraction that interests the viewer. By means of contrast, differences become much more noticeable. A contrast arises when at least two opposing effects coincide. A simple example is a mown path through a flowering meadow. In natural landscapes, many such examples can be found. A beech wood in spring has a highly visible ground covering of wood anemones (Anemone nemorosa), which contrast with the wide, bare tree trunks.

The deliberate association of plants with contrasting forms, sizes and colors is an important tool in emphasizing the impact of individual plants. Strong contrasts, e.g. color contrasts, are perceived quickly, with little concentration required. Weak contrasts, e.g. texture contrasts, require a longer observation time and a more intense concentration on the plants. Contrasts in displays can surprise the viewer, when they are arranged in such a way that he or she is confronted by a new effect on turning each corner.

Contrasts require balance. A quiet background, such as a building's wall or cold and neutral plant colors (green, gray), or a transition mediated by displays graduated by height and color accentuate their contrasting counterpart. Small plants and plants with subdued colors should be planted in greater numbers than larger plants and those with brilliant colors. Too many strong contrasts have an exhausting effect, whereas too much similarity and a lack of clarity appear unsatisfactory and dull. Contrasting pairs that are suitable for design using plants include:

_ Growth form contrasts
_ Texture contrasts
_ Color contrasts
_ Light-dark contrasts
_ Figure-background contrasts
_ Fullness and emptiness
_ Light-shadow contrasts
_ Negative-positive contrasts (concave/convex)
_ "Yin and Yang"

Growth form contrasts

Growth form contrasts heighten the static and dynamic effect of plantings. By enlisting its opposing pole, the growth form of a plant with its own peculiar qualities can be expressed more strongly than is possible in isolation. › Fig. 73 Contrasting pairs are only perceived as such when they are of equivalent size. › Chapter Spatial structures, Proportion Suitable growth form contrasts include:

✎
\\ Tip:
In natural landscapes there are numerous examples of relationships between plants, which may be used as a model for new ideas, e.g. birch woods with ferns. Continually observing and analyzing natural landscapes during walks, rambles and excursions helps to develop a sensibility for design involving plants.

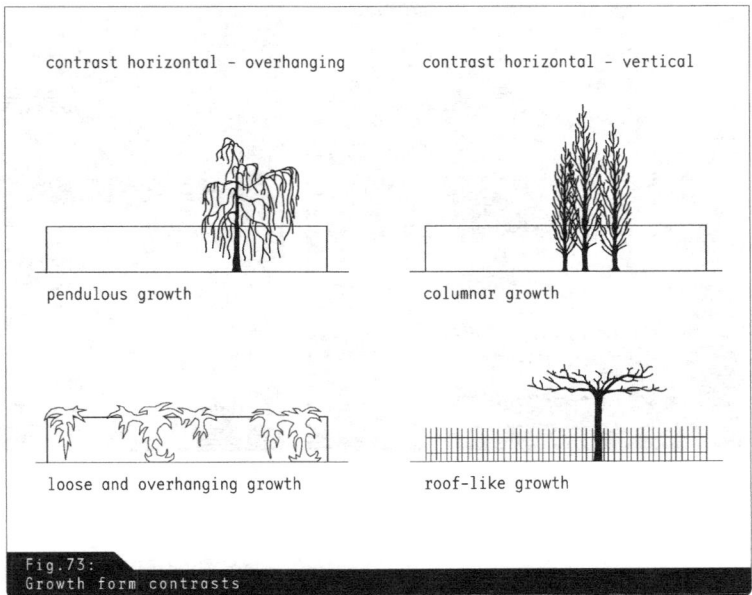

contrast horizontal - overhanging   contrast horizontal - vertical

pendulous growth   columnar growth

loose and overhanging growth   roof-like growth

Fig.73:
Growth form contrasts

_ Vertical and round, without direction
_ Horizontal and loose
_ Overhanging and inclining
_ Loose and firm, round
_ Loose and strict
_ Linear and without direction, round
_ Linear and flat
_ Graphic and artistic

As the sphere form is directionless and has a static effect, a contrast to it can be created using flowing forms with a non-constant direction. A curved ribbon of plants, or spherical plants flanking a winding path may fulfill this function. This motif appears in the natural landscape in the form of erratic boulders and river gravel located in winding riverbeds. Linear leaf forms (e.g. grasses or irises) create a contrast to broad, round and flat leaves (hosta, water lily). Horizontally oriented plants with horizontal branches and a broad umbrella crown (Indian bean tree, catalpa) or clipped hedges offer a recumbent contrast to curved ground forms and surfaces or vertical forms (pillar-shaped woody plants, buildings). A ground-covering planting of a low herbaceous plant variety may also create a simple but

73

Fig.74:
Growth form contrast

Fig.75:
Textural contrast

effective growth form contrast with vertical tree trunks. › Fig. 74 Vertical forms always appear to be nearer than the usually distant horizon line. For this reason, pillar forms in a landscape are striking even when seen from a distance. On curved ground, vertical forms appear fixed by comparison; balancing this using plants with non-constant direction (e.g. inclining or overhanging plants) brings a dynamic into the display. Trees with compact, continuous contours and trees with a graphic, linear effect, for instance, create a juxtaposition that is effective in design terms. › Chapter Plants as a material, Plants: appearance

Textural
contrasts

Textural contrasts lend creative force to a plant schema. This is especially clear in a planting that is quiet in terms of color. In a layout with varying levels of green, the viewer's attention is directed to the interplay of the contrasting foliage and the way the plants are formed. › Fig. 75 White blossoms or white-edged or variegated leaves may heighten the effects of textural contrasts, as they do not detract through colorfulness. Textural contrasts in plants may be:

_ Loose and dense
_ Fine and coarse
_ Lustrous and matte
_ Soft and firm
_ Felted and smooth
_ Rough and smooth
_ Delicate and tough

_ Transparent and leathery
_ Linear and broad
_ Linear and directionless

Coarse-textured plants give an impression of strength and stability, while fine-textured plants radiate peace and understatement. Seen from an equal distance, large-leaved plats appear to be nearer to the observer than plants of the same size with a fine texture. › Chapter Plants as a material, Plants: appearance

Color contrasts Color contrasts make displays livelier and heighten the effect of color. The most important color contrast effects are:

_ Light-dark contrasts
_ Cold-warm contrasts
_ Complementary contrasts (opposing colors in color wheel)
_ Quality contrasts (color contrast of brilliant and dull with texture contrast of lustrous and dull)
_ Quantity contrast (color surfaces of different sizes)

The strongest color contrasts are achieved through the use of two-part and three-part harmonies, that is, using colors that are opposite each other on the color wheel (complementary colors). › Chapter Plants as a material, Plants: appearance Flower colors should harmonize with each other as well as with the surrounding leaves (the base color). Leaves vary in their shooting, summer and autumn color, and also depending on the plant species (yellow-green, green, blue-green, red-brown etc.).

As an alternative base color to green, most spectral colors can be combined with silver-gray to good effect. Colors such as red, yellow and blue gain in strength in this combination, and pink and pastel tones are

＼＼Important:
The principle of "less is more" also applies to design using color. It is reduction that clarifies the design idea and turns it into a strong statement. To choose a single plant species but use several varieties of it is one possibility for simple, powerful design. For instance, the iris has perfectly formed, beautiful flowers with a wide range of colors and simple sword-shaped leaves.

brought out fully. Earth colors present quiet, effective combinations with silver-gray. Gray-leaved plants are generally used in small-format displays, together with small willow species or lavender, for instance. The color white can be combined with others without problems. It heightens all other colors. Equally, various color compositions can be spatially neighbored by white. White flowers have an enlivening, freshening, delicate character. However, plants with white blossoms are more effective in a bloc (a white garden) than when planted in a bed with many different colors of blossom. Planted in front of dark conifers or in a shady area, woody plants with white blossoms create a good light-dark contrast, just as the white trunks of birches do when placed before a dark, flat background. › Fig. 76

The species, amount and distribution of the color tones used must be balanced. Low-brilliance plants will outweigh high-brilliance plants visually only if the former are used in greater quantities than the latter. In Goethe's schema, the brilliance of colors is expressed in "light values":

Yellow	= 9
Orange	= 8
Red	= 6
Green	= 6
Blue	= 4
Violet	= 3

These light values can be used to measure color component amounts: yellow and violet (9/3) = 1:3 or blue and red (4/6) = 3:2. The larger the display to be designed, the more influence the viewing distance has on the planned size of the expanses of color.

Color contrasts may be implemented within a bed, but can also be effected by beds opposite one another but both kept to a single color tone.

Light-shadow contrast

The play of light and shadows on trees and the play of shadows on the ground are very attractive. Depending on its intensity, light creates strong graduations of light and dark. Depending on the color tones of leaves, bark and soil and the nature of the foliage and branch/twig structure, a unique pattern of shade is created: shot through with light, light, dark, heavy, sharp, soft, colorful, full of contrasts, diffuse. Beneath trees, the shadows on the ground change continually. The form of the shadow tells us the time of day. At midday, sunlight is bright and hard, and shadows are short, while in the late afternoon the light is soft and yellow, and the shadows grow increasingly long, strengthening the impression of three-dimensionality in an open space. If the viewer looks into light or at the sun, he or she will blink, while from the shade the landscape can be viewed at leisure. Depending on the season, we seek out either sun or shade. In winter, we love the warmth-bringing sunbeams, while in summer we love the cooler,

Fig.76:
Light-dark contrast

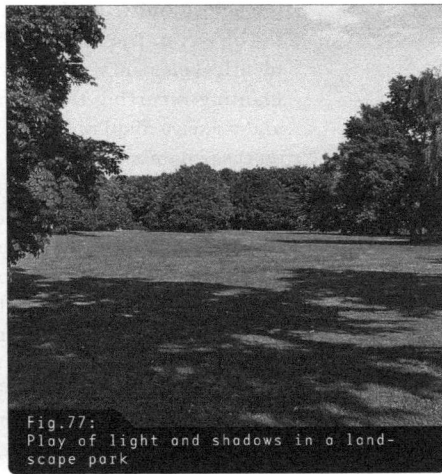

Fig.77:
Play of light and shadows in a landscape park

protective shadows under trees. It is important to know the effect and significance of shadows in an open space, and to offer the appropriate options
when designing an open space.

The play of light and shadows gives trees body and an artistic appearance. The influence of light creates a shadow from the tree's outline.
Loose groups of trees that are lit up by the sun on one side, and in shadow
or casting shadow on the other are a captivating sight. On the other hand,
a continuous edge to a wood or grove appears dull. Incisions in this edge
or displays planted in front of it create expanses of sun, light and shadow
that structure the edge, lighten it and offer a view to the observer. › Fig. 77

Rhythm

In order to give a garden, a park or a display coherence and structure,
it is necessary to repeatedly include the same or similar plants or groups
of plants. Simple repetition alone does not create rhythm, but only a connection between certain areas of a display. Rhythm, and with it a unified
overall layout, arises with regularly recurring characteristic vegetation elements; close and distant areas are connected with one another visually. In
a garden or park, an atmospheric unity of the whole layout can be created
by uniting its individual areas by rhythmical elements of different characters. If a typical plant species is used in an open space layout in large
numbers, it characterizes this space unmistakably, becoming a theme; one
thinks of the chestnut avenue or the rose garden.

Theme plants

In a crowd of equally prominent but diverse plants, our eyes cannot orientate, and our gaze passes over them. The impression created is

77

undecided, inharmonious and lacking in interest, and does not appeal to the observer. The human eye will more readily perceive several similar or identical elements, as they are more easily readable and create a structure. Planting recurring theme plants or plant groups creates visual stability and makes a display comprehensible. The gaze can always return and light on these points or surfaces.

Theme plants create a starting point for planning a display; a framework. Their arrangement holds the display together. Plants are placed, grouped and repeated according to their ranking. Plantings of woody plants are defined most strongly by trees. Trees form a lasting framework and create a connection with the rest of the settled area or with the landscape. They may stand, for instance, near a building, in particularly architectonic parts of a garden, at the corners of plots or in edge plantings. The size of trees should harmonize with the architecture and the space available. › Chapter Spatial structures, Proportion Small trees, large bushes and solitaire woody plants are, like trees, structuring plants. They serve to emphasize size relationships, connect architecture with the rest of the garden space and create a transition to bushes. They may stand, for instance, at the entrance of a garden plot, at the corners of houses or outbuildings and in edge plantings. Bushes and hedges are secondary to the structuring plants, and serve as spatial termination and demarcation for the different parts of a garden. Together with the theme woody plants, they can lend character to a garden space. The subsequent planting of small bushes and dwarf woody plants fills out the profile of such a woody plant display. › Tab. 6

Naturalistic expanses of herbaceous plants are structured using solitaire and theme herbaceous plants. Theme herbaceous plants are also described as core or framework herbaceous plants, as their effect is appealing throughout the growing period of a year, and they are long-lived. Tall species whose form or color is apposite and effective are used. Additional

\\Tip:
Plant combinations that appeal can be noted and sketched out in a small notebook, which should always be kept to hand. Through thoughtful observation, noting and drawing, the image of the plants is imprinted more firmly on the memory, and a fund of personal knowledge is created, which one can refer to in one's own plans.

\\Tip:
The book *Perennials and Their Garden Habitats* by Richard Hansen and Friedrich Stahl is a standard reference work on the use of herbaceous plants (see Appendix, Literature) and is useful in choosing plants with reference to growing zones and habitat requirements, as well as calculating the number of plants required per square meter, intervals between plants, and sociability.

Tab.6: Classification of planting of woody plants		
I a	Trees	Primary woody plants. Lasting framework plants. Require adequate growth space. Isolated or in groups
I b	Small trees, large bushes, solitaire trees	Framework plants. Larger than neighboring plants. Long-lived. Isolated or in small groups. Take first place in ranking system in smaller gardens. These determine choice of bushes.
II	Bushes, hedges	Accompanying woody plants. Significantly smaller maximum size than primary woody plants. Suitable bushes may form a framework or flowering accompaniment for herbaceous planting. Isolated or in small thickets
III	Small hedges, dwarf woody plants, semi-bushes	Complementary woody plants. Smaller maximum size than trees and hedges. Use as ground cover under taller-growing woody plants. May form the framework for a herbaceous display

Tab.7: Classification of herbaceous planting		
I	Solitaire herbaceous plants	Effective. Only a few individual plants required
II	Theme herbaceous plants	Framework herbaceous plants. Larger than neighboring herbaceous plants. Long-lived. Appear in greater numbers
III	Accompanying herbaceous plants	Support the theme herbaceous plants in their effect. Smaller maximum size than framework herbaceous plants. Long-lived
IV	Filling plants	Smaller maximum size than theme and accompanying herbaceous plants. Their growth should not be allowed to affect the theme and accompanying herbaceous plants.

herbaceous plants accompany the theme herbaceous plants rhythmically in greater numbers and should therefore be more modest in appearance. Filling herbaceous plants are used to create surfaces or to cover the ground. The transitions between the different types are fluid. One and the same plant type can occupy a different position depending on the theme of the garden. For instance, iris may be the theme plant for one theme, and the accompanying plant for another. > Tab. 7

Height tiering

One possibility when arranging herbaceous plants is a three-tiered structure of tall, medium and small species. The rhythmical repetition of herbaceous plants should not be schematic, as this causes the display to lose interest and liveliness. The intervals between the theme plants and the number of individual theme plants should be varied, with the relative expanse of each of the tiers changing along its length.

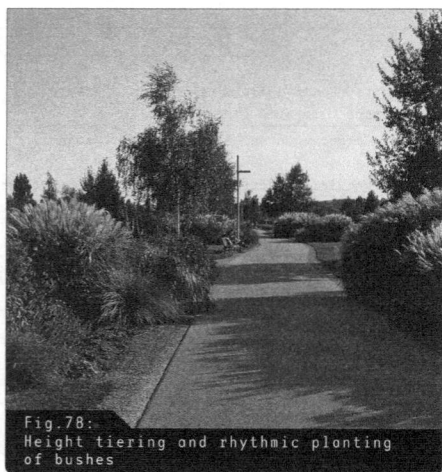

Fig.78:
Height tiering and rhythmic planting
of bushes

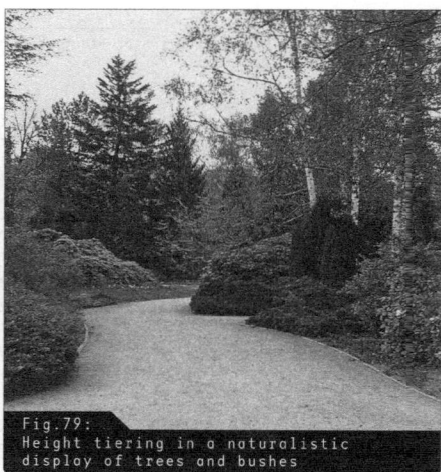

Fig.79:
Height tiering in a naturalistic
display of trees and bushes

Fig.80:
Height tiering in a formal rose garden
with a three-tiered structure

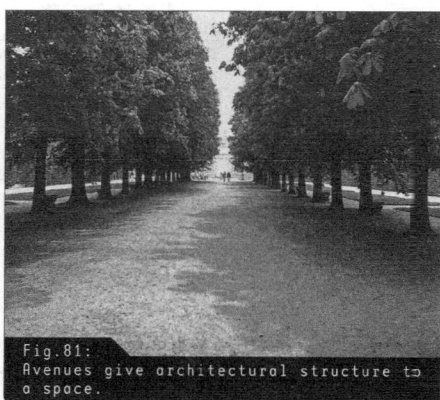

Fig.81:
Avenues give architectural structure to
a space.

The lower plants in a bed can extend further forwards or less distance backwards. Intermediate or tall plants can be placed forward, move forward or recede. › Fig. 78 An even distribution of lower plants in the forward part of the bed, medium plants in the middle and tall plants in the rearward expanse (or, in a bed that can be viewed from all sides, in the middle) has a dull, lifeless effect. A further compositional possibility is a two-level construction. This involves placing taller plants individually or in small groups amid surface-covering lower varieties. Displays of woody

Fig.82:
Yew cubes arranged in a grid

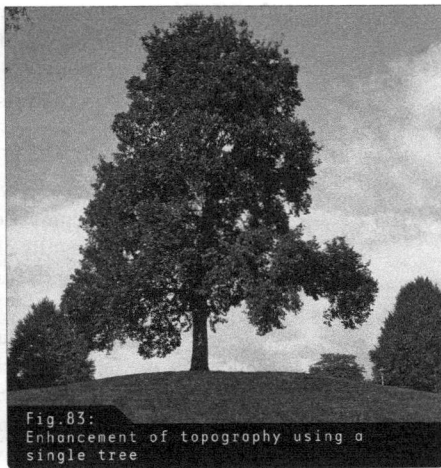

Fig.83:
Enhancement of topography using a
single tree

plants and combinations of woody plants and herbaceous plants are also tiered according to an established ranking system. > Figs 79 and 80

Repetition and
heightening

The simplest form of repetition is to position identical elements at regular intervals. This creates a clear continuity with a high degree of unity. For instance, trees may form a row, avenue or tree block. > Fig. 81 The effect of these elements is strict and formal. The resulting regular arrangement can be extended as far as is desired.

Through repetition, a selected plant is emphasized and its significance strengthened. The repeated element may be the intervals between plants (e.g. a grid pattern), their color or texture. > Fig. 82 The heightening of a plant theme can achieve an even more expressive overall effect by graduating flower color, sizes and textures. This involves using different sorts of the same plant or alternations of accompanying plants. Important: The choice of species, in particular the theme herbaceous plants, should be restricted, as above all good design with plants involves clarity and simplicity.

A further form of heightening involves accentuating already existing constructed or topographical features using plants. For instance, a regular tree block echoes the orthogonal form of a building, a group of trees emphasizes a hill, or an avenue of trees accompanies a street. > Fig. 83

Symmetry and
asymmetry

The reflection of an individual plant, shaped plants or area figures in an axis creates a symmetrical effect. Paths may form axes of symmetry. Possibilities for planting include loggias, trellises, pergolas and trees with repeated form and characteristics (e.g. avenues and shaped woody

81

plants). Trees standing in pairs indicate the border to a space, a change in function in the course of a path, or a construction relating to it such as an entrance gate, bridge or steps. A symmetrical arrangement can be repeated several times in its entirety (e.g. as an ornament). › Fig. 84 Planted parterres and sightlines in Baroque gardens are typical examples of planning using axes of symmetry. In such a layout, ornamental area figures are surrounded by formally clipped hedges, in order to reinforce the impression of being symmetrical. Symmetrically arranged areas can also be surrounded with formally clipped hedges, behind which trees and bushes grow naturally. Due to the way landscapes and parks are perceived by the observer when passing through, fleeting natural symmetrical effects can be created by blocks of plants continually receding into the landscape in such a way that one block occasionally appears to be the same size as another block opposite to it. The creation of true symmetrical effects restricts the designer. This kind of design can be used in formal and imposing gardens in connection with built elements or ornamental beds, such as bedding plant displays. › Chapter Plants as a material, Time dynamics

Equilibrium

Equilibrium is a common aim of design. It describes a state of balance and harmony between different design components. We experience a balanced design as harmonious and not as stiff as a symmetrical framework. Balance and symmetry may be achieved together within a landscape, park or garden by means of a central built element. The exact positioning of plants on each side creates symmetry, while small variations in planting create balance. › Fig. 85 The more the building obtrudes visually, the less it is necessary to resort to symmetry in the planting. One possibility is to place plants with striking forms, textures or colors at appointed intervals on either side of an axis of symmetry, while structuring the planting in between less strictly.

Pictorial
design

Pictorial design involves working with dissimilar visual objects, the intervals between which are irregular. › Fig. 86 Both free and geometrical forms may be used, and may be combined. The arrangement of the plants is usually more important than the plants themselves. Intervals between plants and plant forms must be chosen and positioned in a balanced way. The visual focus lies outside the centre of the surface.

Fig.84:
Symmetry

Fig.85:
Asymmetry

Fig.86:
Pictorial design

IN CONCLUSION

The fascination of garden design lies in the ambivalence between the static and the living, the fluid identity of plant and space. Everything living is influenced by time and space. Landscape design using plants is a form of artistic expression that, perhaps more than any other, is dependent on intensive observation of time and space. Planting a garden is the beginning of an ongoing process. The design and creation of gardens is inseparably linked to horticulture, as only active cultivation can ensure that the designer's vision is enabled to develop. Utilizing plants requires gardening knowledge. It is not a question of either architectural plantings or wild-growing hedges. There is an abundance of other possibilities, such as English-style herbaceous plantings or recent examples from Scandinavia from which inspiration can be drawn. As when choosing materials, the following applies: a passion for reduction should not cause us to forget that reduction begins with abundance, from which it makes a considered choice. If there is only a small choice available at the beginning, the impression is one of poverty rather than reduction. Plant-based design in our gardens will continue to gain in importance. As a place of work, leisure and recreation and a sign of a degree of expendable resources, gardens create a world in contrast to an increasingly mechanized, dependent society. The need to care for plants, rather than the ease of caring for them, will return to the foreground. A sharper consciousness of beauty goes hand in hand with this, awakening our senses. Design using plants is a great luxury in our time, because it demands those things that are most rare and costly in our society: time, attention and space. Use of plants signifies our perception of nature. When we re-incorporate intellect, knowledge and craft a responsible way of working with the environment and its microcosm, the garden, is created.

APPENDIX

PLANTING PLAN

Planting plans show the species of intended plants, their location and number schematically. If a plan is drawn to scale, it is possible to calculate the actual number of plants needed for a layout in advance and to select combinations of plants which are suitable, not only from a planting point of view, but also for the proportions of the open space being laid out. > Fig. 87, p. 86, and Fig. 7, p. 19 A planting plan is a tool that clarifies the development of the planting idea step by step for the planner, initially on drawing paper or on a computer screen. In a site plan, trees should be represented with trunks and crowns, so that it can be seen that they have an impact on the design of both spaces and surfaces. The crown of the tree is its body, the stem of the tree marks its location. The extent of the roots is generally the same as the crown, and should stop short of the area around built structures (buildings, underground pipes, roads). > Fig. 88 A planting plan provides the supervisor onsite and others involved with the information they need to realize the design idea. The types of plant and their locations can be assessed quantitatively and in design terms, area covered by plants and turf, necessary anchoring, soil improvement measures etc. can be investigated mathematically, landscape gardening work can be calculated and the course of the work prepared.

\\ Tip:
Lists and tables in the appendices of tree nursery and herbaceous plant catalogues offer information on growth forms and the number of individual plants required per m^2 (see Tab. 8).

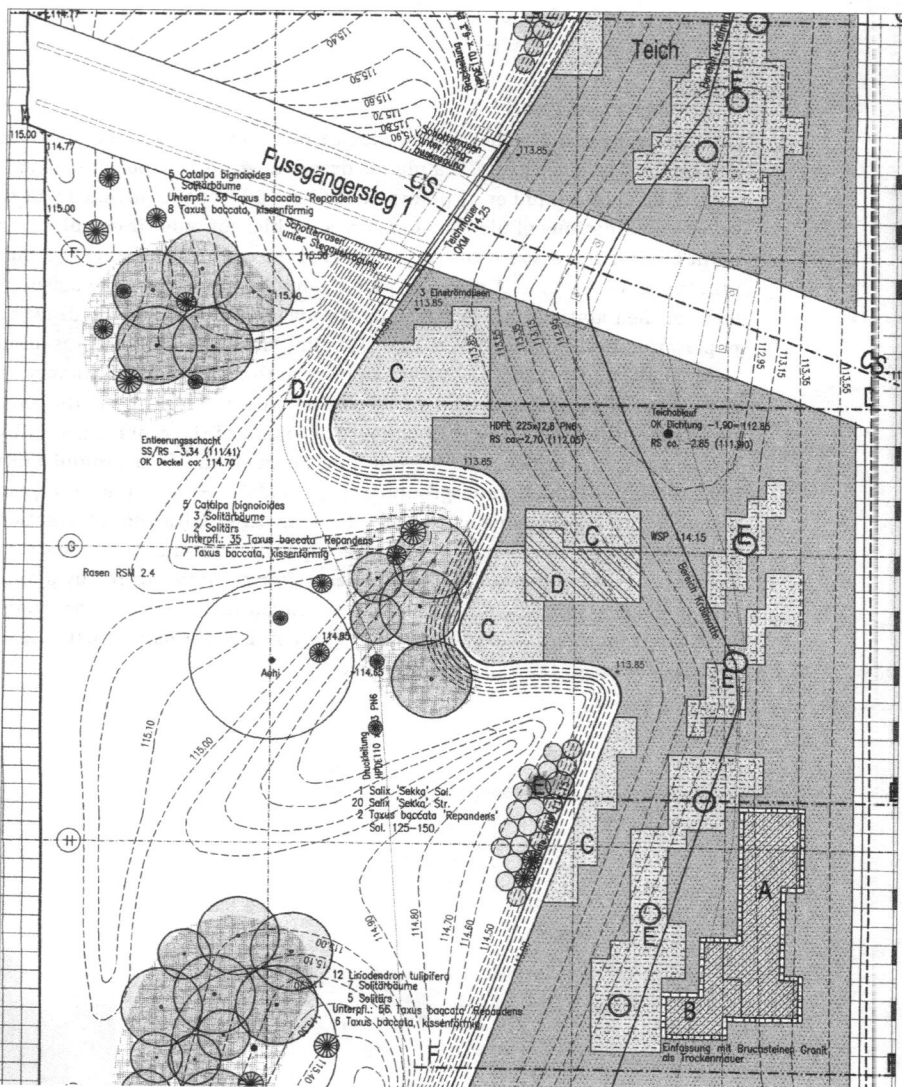

Woody plants

- ⊙ Tree stand
- ● Tree planning
- ░ Taxus baccata formally clipped yew "cushion"
- ✳ Under-planting Taxus baccata "Repandens"

Pond plants

- C Large-leoved herbaceous plants. Medium-leaved and long-stemmed herbaceous plants
- ▦ V-B Vertical herbaceous plants
- ⊚ E Reed-type herbaceous plants
- ▨ D Shaft ring with horizontal herbaceous plants with buyoant leaves

Fig.87:
A specimen planting plan

crown

trunk

root space

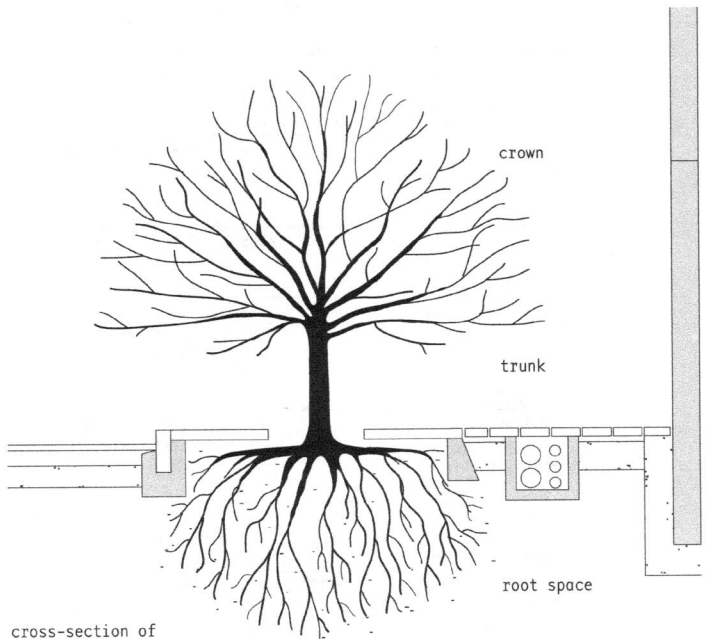

cross-section of
tree's circumference

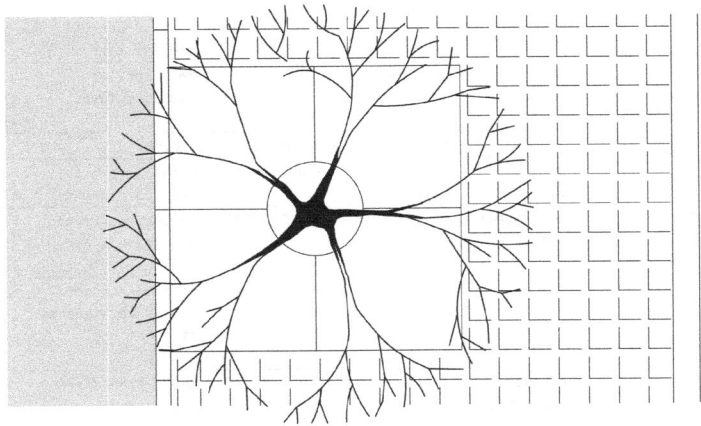

top view of tree's circumference

Fig.88:
The whole structure of a tree must be taken into account during planning.

Botanical name	English name	Height in meters	Breadth in meters	Characteristics/ form	Peculiarities
Small trees for gardens and urban spaces					
Acer campestre "Elsrijk"	Field maple	8–12	4–6	Compact, conical	Beautiful autumn coloration (yellow), tolerant of urban climate
Acer plata-noides "Globosum"	Norway maple	4–6	3–5	Compact, spherical, loses proportions with age	Autumn coloration (yellow), tolerant of urban climate
Amelanchier lamarckii	Juneberry	5–8	3–5	Bush-like, broad-growing, funnel-shaped	White flowering racemes towards the end of April, beautiful autumn coloration (yellow to flame red)
Carpinus betulus "Fastigiata"	European hornbeam	10–12	5–8	Pillar-shaped, stiffly upright	Has a narrow crown even when unclipped, growth remains tight as tree ages
Catalpa bignonioides "Nana"	Indian bean tree	4–6	3–5	Dense spherical shape	Beautiful large leaves, slow-growing, no blossoms
Pyrus calleryana "Chanticleer"	Callery pear	7–12	4–5	Regularly spherical	Tolerant of urban climate, very tolerant of heat, white blossoms, beautiful autumn coloration (scarlet)
Sorbus aria	Whitebeam	6–12	4–8	A large bush with multiple trunks or a broad conical smaller tree	Orange-red decorative fruit after September
Tilia europaea "Pallida"	Common lime	Formal cut		Box, roof or espalier form	Clipped woody plant crown forms a body with handsome volume
Medium to large-crowned trees for towns and parks					
Acer platanoides	Norway maple	20–30	10–15	Large tree with rounded crown	Tolerant of urban climate, fast-growing
Acer pseudo-platanus	Sycamore maple	20–30	12–15	Extensive, broad and round crown	Fast-growing, autumn coloration golden-yellow
Aesculus x carnea "Briotii"	Red horse-chestnut	8–15	6–10	Rounded compact crown, main shoots straight and upright	Slow-growing, brilliant red flower panicles, forms hardly any fruits
Aesculus hippocastanum	Horse-chestnut	20–25	12–15	Oval, high-domed dense crown providing a high degree of shade	White blossoms, produces many fruits, beautiful autumn coloration

Ailanthus altissima	Tree-of-heaven	18–25	8–15	Oval broad round-crowned tree	Fast-growing, unde-manding and tolerant of urban climate
Betula pendula	Silver birch	12–25	6–8	Elongated egg-shape with loose over-hanging twigs	Catkins yellow-green, bark whitish-brown, autumn coloration
Catalpa bignonioides	Indian bean tree	8–12	5–8	Umbrella-shaped domed crown	Large heart-shaped leaves, impressive 15–30 cm long flower panicles
Corylus colurna	Turkish hazel	12–15	6–8	Cone-shaped crown, continuous main shoot	Tolerant of urban climate, robust, undemanding tree
Fagus sylvatica	Copper beech	25–35	15–20	Very extensive oval crown	Silver-grey trunk, autumn coloration yellow to orange
Fraxinus excelsior	European ash	25–35	15–20	Egg-shaped crown, becomes extensive with age, admits dappled light	Beautiful pinnate leaves, autumn colora-tion rare
Platanus acerifolia	Hybrid plane	25–35	15–25	Broad conical round-crowned larger tree, becoming extensive with age	Very vigorous growth, tolerant of clipping, tolerant of urban climate
Populus nigra "Italica"	Black (Lombardy) poplar	25–30	2–5	Pillar-shaped large tree, branches and twigs straight and upright	Strong growth, tolerant of flooding
Prunus avium	Wild cherry	15–20	8–12	Egg-shaped crown of medium size	Very beautiful tree with white blossom, glorious autumn coloration (yellow to orange)
Quercus robur	Pedun-culate (English) oak	30–40	15–25	Initially cone-shaped crown, becomes extensive, loose and round with age	Tolerant of urban climate, wind-resistant
Salix alba "Tristis"	White willow	15–20	12–15	Dramatic ornamental tree of medium size. Drooping overhanging branches	Artistic charac-teristics. Becomes vulnerable to wind damage with age
Tilia cordata	Small-leaved lime	20–30	10–15	Grand large tree. Crown initially cone-shaped, later high-domed	Moderately tolerant of urban climate, tolerant of clipping
Pinus sylvestris	Scots pine	15–30	8–10	Artistic large tree with a variable form, develops a high crown with age, umbrella-shaped	Paired needles, green to blue-green, tolerant of urban climate
Thuja occi-dentalis "Columna"	White cedar	15–20	2–3	Pillar-shaped tree of medium height	Evergreen, tolerant of urban climate, tolerant of clipping

LITERATURE

Ethne Clarke: *Gardening with Foliage, Form and Texture*, David & Charles PLC, Devon 2004

Rick Darke: *The American Woodland Garden*, Frances Lincoln, London 2000

Brian Hackett: *Planting Design*, McGraw-Hill, New York 1979

Richard Hansen, Friedrich Stahl: *Perennials and Their Garden Habitats*, Cambridge University Press, Cambridge 1993

Penelope Hobhouse: *Colour in Your Garden*, Collins, London 1985

Penelope Hobhouse: *Penelope Hobhouse's Garden Designs*, Frances Lincoln, London 2000

Gertrude Jekyll: *Getrude Jekyll's Colour Schemes for the Flower Garden*, Frances Lincoln, London 2006

Noël Kingsbury: *Gardens by Design*, Timber Press, Portland, OR 2005

Hans Loidl, Stefan Bernard: *Opening Spaces*, Birkhäuser Verlag, Basel 2003

Piet Oudolf, Noël Kingsbury: *Designing with Plants*, Conran Octopus, London 1999

Piet Oudolf, Noël Kingsbury: *Planting Design: Gardens in Time and Space*, Timber Press, Portland, OR 2005

Marco Valdivia, Patrick Taylor: *The Wirtz Gardens*, Exhibitions International, Leuven 2004

James Van Sweden, Wolfgang Oehme: *Bold Romantic Gardens: The New World Landscape of Oehme and Van Sweden*, HarperCollins Design International, New York 2003

Rosemary Verey: *Rosemary Verey's Making of a Garden*, Frances Lincoln, London 2006

PICTURE CREDITS

THE AUTHORS

Regine Ellen Wöhrle and Hans-Jörg Wöhrle, Dipl.-Ing., practicing landscape architects and proprietors of the w+p Landschaften landscape architecture practice in Berlin, Stuttgart and Schiltach in der Kinzig.

前言

　　空间与植物——即利用植物塑造空间与景观建筑学——长久以来一直都是园艺史中最为重要的内容。然而，好的设计并不仅仅是好看的问题。很多情况下，好的设计都是与对应环境的需求相吻合的。植物设计不仅需要有好的设计能力，更多时候需要的是知识。也就是说，我们要学会利用感官感受的法则去有效地体现设计意图，并获得他人的认可［希腊语"aesthesis（美学）"一词就是指有关感官感受的科学］。由于没有任何两块场地是完全一样的，因此任何现成的设计和范例都不能简单地照搬，这一事实更加突出了知识的重要性。另一方面，一些关于设计、空间、序列、对比、平衡、重复等普遍适用的基本原则可以用在任意场地的设计中，并最终辅助我们得到一个好的设计结果。

　　植物是一种可以用来塑造室外空间的有生命的天然材料，它与我们都市文明中日益增长的技术环境形成了鲜明的对比和反差。由于植物能够营造氛围，因此树、灌木以及草本植物可以在文明与自然之间的灰色地带营造出多种多样的室外空间。植物设计会遇到很多不同的情况，从为私家住宅做设计到为大型建筑物设计庄重的场地环境，再到为一个城市去设计包含公共空间、步行空间、公园、通向城市的绿化道路、休闲娱乐区、墓地以及认养绿地在内的复杂的绿色空间体系等。

　　本书可以培训我们在室外植物设计上的能力，并让我们逐渐认识到，只有在设计的过程中结合植物给予统筹考虑，才能将建筑设计以及城市规划设计的想法提升成为一个统一的、高层次的理念。

基本设计原则

暂且抛开具体内容不论，我们在动手设计之初或是在着手实现某个想法之前，下文列举的几个关键要素都是需要给予了解和考虑的。对于这些要素，我们需要建立基本的认识和判断。

植物的生态栖息环境

在野外旅行的过程中我们会发现景观是在不断发生变化的；伴随着地势的升高和温度的降低，土壤肥沃的温带水果和葡萄生长区就会逐渐被生长在贫瘠土壤上的落叶林和针叶林所取代。生长在温带气候区的植物耐寒性差，如不采取保护措施，它们在冬季极易受到伤害。在适宜的环境和肥沃的土壤中长势良好的植物在贫瘠的土地上生长时会变得矮小。因此，要想做出好的植物设计，就必须了解这些植物的生长环境和视觉效果。〉见"植物也是一种材料，植物：外观"一章影响植物生长的因素有很多；〉见图1、图2

—气候；

—土壤；

—pH 值；

—地理位置；

图1：
气候决定了植物的生长区域及分布范围

图2：
光是环境因素，土质是地理因素

— 光照；

— 水分；

— 养分；

— 竞争。

气候

任何一块场地都会受到其所处的大环境及海拔高度的影响。这些先决的自然条件是无法改变或规避的，它们决定了不同植物的分布范围。但是，生长区域的小环境却是可以改变的。墙的内拐角或是建筑的内转角（例如建筑的内庭院）等区域就不会受到风和辐射热的影响。如果同时还具备其他有利的生长条件（如土壤、降水等），那么适合在这个区域生长的植物种类就会有所增加。下述的气候参数对植物的生长十分重要：

注意：

生态学是研究生物有机体之间的相互关系、相互作用以及生物有机体对其生存环境的适应性的科学。研究环境对个体植物行为的影响以及环境因素对植被组群影响的学科称为植物生态学。

—温度：冬季低温及夏季高温；

—湿度：夏季降水量及冬季降水量。

最为重要的一个因素是冬季的寒冷程度，某类植物能否在冬季存活就取决于冬天的最低气温。耐寒度是指某种植物在不受霜冻侵害时能够存活的最低温度。对于夏季的温度而言，极端的高温并不重要，而温度的总量——夏季的平均温度——才是决定性因素。只有达到特定的温度，植物才能发芽、长叶、开花、结果。气候越是温和，可供选择的植物种类也就越多。

土壤、位置、pH 值

土壤为植物的生长提供了支持。植物从土壤中获取水分和养分并扎根于土壤之中。土壤的构成以及水分和养分的含量等因素对于植物的生长都至关重要。在为场地选择植物品种时，还要考虑现场土壤的类别（黏土、亚黏土、沙土、粉砂土）和 pH 值（酸碱度）等因素。不同的植物对 pH 值的要求也不尽相同。而 pH 值又决定了土壤养分的含量；酸性土壤养分含量低，而碱性土壤养分充足。

区域小环境会因场地的坡向和日照程度的不同而有所差异。朝南的坡地相对比较温暖和干燥。朝北的坡地则会相对寒冷和湿润。人们在穿越山区的过程中会发现，山的北坡草地和南坡草地上生长的花卉在品种上有着很大差异，而花的色彩构成也因此大不相同。

光照

生长环境的日照程度则决定了一株植物能否在此地存活并茁壮生长。根据日照情况，场地可分为"完全日照区"、"直接日照区"、"间接日照区"、"部分阴影区"、"全阴影区"等几种类别。有些植物喜光但不耐阴，有些植物可以同时适应这两种条件，例如雪果（snowberry）。由于植物的生长会使植株的外形和间距发生相应的变化，因此随着时间的推移，植物接受日照的时间也会有所改变。关于树阴还会在后面特别予以介绍。见"植物也是一种材料，时间力学"一章日照时间不仅是选择单株植物时需要考虑的一个因素，它也会影响到整片绿植区或是花园局部区域的品质。对于明显的阴影区和光照区而言，这一点体现得尤为突出。由于在阴影区内开花现象会大量减少

注意：

土壤的成分是可以改变的。但是需要长期的维护，否则现有的自然条件迟早会让一切都回到从前。

因此该区域的主要特点就集中体现在木本植物及草本植物的叶片形态、颜色和质感上。

水

水是植物最为重要的组成原料和油料。因此在自然环境中，降水量的多少至关重要。特别是在夏季，降水可以让生长中的植物避免在遭遇高温和高强度光照时干涸死亡。在冬季，大多数植物都会因叶片的脱落而处于休眠状态，因此对水的需求量很小。冬季降水（雪）对于耐寒植物非常重要，因为厚厚的雪层可以让植物靠近地面的部分和地面以下的部分免受严寒的侵害。如果没有雪层的覆盖，严寒会给常绿植物造成伤害。随着水分从植物叶片逐渐蒸发出去，而植物又无法从坚硬的冻土层中汲取水分，就会因此出现脱水问题（霜冻干旱）。

土壤的自然含水率受降水量、当地地下水位的高度、土壤的构成和渗透性以及场地坡度等因素所决定。植物能够汲取到多少水分取决于植被的外形结构以及由于物种竞争所造成的该结构体系的破坏程度。但是，不同的植物对于水分的需求是有差异的。有些植物偏好干旱的环境，而有些植物只有生长在水中才能枝繁叶茂。例如，在自然环境中，松树和金雀花（gorse）在低矮的草丛陪伴下可以孤零零地生长在日照充足的透水沙土中。它们属于耐旱植物。它们的叶片呈针状，非常坚硬，又细又长，因此可以适应严酷的环境条件。

注意:
　　均匀布置的喷灌设施能够改善种植区的环境，但是也会增加维护的成本。地面过多的积水，例如密实场地表面的积水，可以通过排水设施予以排除。

竞争

在自然环境中，许多植物都有着类似的生态需求。在这样的竞争环境中，弱小的物种经常会被其他物种所取代，而竞争者本身也无法获得好的长势。遮天蔽日的高大物种就是引发竞争的一个例子，因为它们妨碍了矮小物种获得充足的光照。在进行植物设计时，还需要考虑时间的因素。花园在种植之初会显得空空荡荡，但是随着时间的推移，植物会越长越茂盛。特别是树冠投射在其下方植物上的阴影将会日渐扩大，也会由此引发对日照、水分和养分的争夺。因此，了解植物的生长行为也非常重要，这其中包括根系的生长发展、植物的形态和尺寸的变化等。

　　至于如何选择适合场地生长的植物的问题、如何能让木本植物与草本植物形成协调统一的效果，而且后期维护又无需投入大量人力等问题，"索引体系"（index system）可以给我们提供一些帮助。木本植物的分类采用了四位数字代码。首位数字代表了植物的生长环境：

　　—湿地及沼泽；
　　—滨水草地及滨水木本植物；
　　—多物种森林及木本植物群；
　　—单一物种森林及木本植物群；
　　—荒野及沙丘；
　　—干性草原木本植物及低湿度森林；
　　—凉爽气候区木本植物及湿润型森林；
　　—高山森林及高山灌木；
　　—景观树篱及装饰性植物。

　　第二位数字代表了最为重要的土壤环境因素，第三位数字代表了诸如日照、温度等地面环境因素，第四位数字代表了植物能够生长所及的尺寸。这些索引数字仅仅是给因地制宜的种植设计提供了一些信息而已，而不是植物的社会学分类体系。许多植物都可以适应多种环境，也因此才会存在为数众多的植物混生区和过渡区。草本植物的分类也同样是依据不同的生长区域采用了四位数字代码体系，该代码不仅包含了植物生态栖息环境的信息，同时也包含了植物的功能信息。首位数字代表植物的生长环境：

　　—林地；
　　—林地边缘；
　　—开阔场地；
　　—河床；
　　—假山；
　　—水岸。

　　第二位数字代表了组别（或功能），第三位数字代表了植物生长所需的条件，第四位数字则对该植物的应用给出了特别说明。

P15　　**植物的使用要求及功能要求**

　　植物对生长环境的要求就决定了目标场地所适用的植物种类。

而植物的选择又是受实用、美观和创新性所决定的。设计之初，我们不仅要了解植物的生长需求，还要牢记设计需要满足的功能要求。此外，还要让客户了解植物的生长需求以及基本的维护要求等。例如，在设计私家花园时，设计师需要了解客户的个人意愿以及对空间的构想，甚至要让客户提出的功能要求成为整个设计的主导。当承接的设计任务是为特殊客户群服务时，例如为医院设计室外场地、为养老院设计花园，或是承接公园、游戏场以及公墓项目的设计时，目标客群的想法和需求都要事先给予充分落实，并得到他们的确认。

　　在设计游戏场地时，切记有些树和灌木是会有可能给人们造成伤害的。孩子们喜欢在灌木种植园里追逐嬉戏，经常会有折树枝、揪树叶，或是采花摘果一类的举动。〉见图3 因此有毒的植物不得用在游戏场地的设计中。〉见表1 此外，场地内还要种上几株大树，以便在夏季给人们提供必要的遮阴。〉见图4 树或灌木掉落在地上的枝杈会被小孩子们拿来当做"游戏道具"，而大一些的孩子则喜欢在大树小树上爬来爬去。对于青年人而言，他们不仅需要清静的空间，也需要能够嬉戏和炫耀的场所。但是对于养老院而言，功能需求就会全然不同。在这种情况下，设计的重点就会集中在要让植物营造出怎样的空间感受，

🔍
示例：
　　种有独立大树的花园中，应配上大片的草本植物花圃。与之匹配的草本植物索引代码是4.3.3。4代表适用于花圃的草本植物，3代表该草本植物来自高山森林或是高纬度开阔地区的林地及林地边缘，第二个3代表喜凉耐阴的草本植物。第四位数字除了表述其他内容之外还给出了植物的群集性方面的说明。4.3.3.7表示该植物适合单株栽植或小范围群植（如秋牡丹 *Anemone japonica*），4.3.3.3 表示该植物具有非扩张性或弱扩张性的生长特点，因此可以与其他植物混植（如落新妇 *Astilbe x arendsii*）。

图3：
利用柔软的柳树围合出游戏区

图4：
在设计游戏场地的绿化时，可以为孩子们设计一些景观

以及怎样才能促进居民交往等问题上。色彩、形态和质感的运用应当手法多变，效果突出。小路两侧应栽植遮阴植物并设有长椅，这样人们就可以非常悠闲地去欣赏植物了。座椅的位置应该非常显眼，但又很舒适地被藤架和草本植物围绕着。

要想获得空间的视觉冲击力，最基本的也最为重要的内容包括，大尺度空间关系（如视线）的把控、现有道路及规划道路与街道和路网的联系、初期地形的塑造等，视觉冲击力的塑造要能与设计所涉及的方方面面的内容全部协调一致。对于大规模的绿化工程，可以分期进行建设。因此，种植以及其他重要节点的深化设计，如花架、水景、座凳、照明等可以留到施工后期再行启动。

例如，公墓的格局就是逐步形成的。通常墓地都会占用很大一片场地，而且不大可能全部都被占满，具体占用比例要看土地的使用密度。必备的道路和水路的交通联系也只是在建设的初期才会需要。但是整个场地（包括由树以及边界绿化组成的植物群）的空间格局应该在这一时期形成，这样一来，当场地全部被占满之后，每个单独区域的施工痕迹几乎看不出来。也就是说，所有构建空间格局的植物在生长发展过程中都会步调一致。这会给人们留下一个协调统一的整体印象。这种渐进模式也同样适用于其他场地的绿化设计（如住区设计、休闲公园设计、体育场馆设计等），我们应当广泛运用这一手法来获得整体协调统一的效果。〉见"植物也是一种材料，时间力学"一章

毒性程度	学名	名称	有毒部位*
强毒类	*Aconitum*（包括所有的品种及变种）	乌头	全株
	Daphne（包括所有的品种及变种）	瑞香	全株
	Taxus（包括所有的品种及变种）	红豆杉	除假果（假种皮）外全株有毒
中毒类	*Buxus sempervirens*（包括所有变种）	黄杨木	全株
	Convalleria majalis	铃兰	全株
	Crocus（包括所有的品种及变种）	番红花	球茎
	Cytisus（包括所有的品种及变种）	金雀花	豆荚
	Digitalis（包括所有变种）	洋地黄	全株
	Euphorbia（包括所有的品种及变种）	大戟	全株，尤其是植物汁液
	Euonymus（包括所有的品种及变种）	桃叶卫矛	种子、树叶、树皮
	Hedera helix	常春藤	全株
	Juniperus（包括所有的品种及变种）	杜松	全株，尤其是树枝的嫩芽
	Laburnum（包括所有的品种及变种）	金链花	全株，尤其是花、嫩枝和根茎
	Lupinus（包括所有变种）	羽扇豆	种子
	Lycium halimifolium	中国枸杞	全株
	Rhododendron（包括所有的品种及变种）	杜鹃	全株
	Robinia pseudoacacia	刺槐（洋槐）	树皮
	Solanum dulcamara	欧白英	特别是浆果
低毒类	*Aesculus*（包括所有的品种及变种）	七叶树	果皮及未成熟的果实
	Fagus sylvatica	欧洲山毛榉	坚果
	Ilex（包括所有的品种及变种）	冬青	浆果
	Ligustrum（包括所有的品种及变种）	女贞	果实
	Lonicera（包括所有的品种及变种）	金银花	果实
	Sambucus（包括所有的品种及变种）	接骨木	除成熟的果实外，其余均有毒
	Sorbus aucuparia	花楸	果实
	Symphoricarpos（包括所有的品种及变种）	雪果或白浆果	果实
	Viburnum（包括所有的品种及变种）	荚莲花（八仙花）	果实

*即便植物有毒部分的毒性非常低也依然会招致严重的投诉。

植物与所处环境的关系

任何场地的设计都是以其所在的特定背景环境为依据的。场地及其周边环境就属于背景环境的一部分，而社区环境与社会文化环境也同样扮演着重要角色。在构思设计方案时，对用地、周边环境、历史以及客户进行广泛深入的了解是非常有帮助的。在分析的过程中，要事先对用地的各项要素之间的系统性、依存性、彼此关系以及除此之外的很多东西进行研究了解。这些要素是设计的主体架构和基础。设计方案可以与这个体系取得协调统一，也可以通过其他方式对其进行阐释。同样，你也可以有意识地去寻求它的对立面，或是提出一个可以脱离于现有体系之外的方案。

关注用地有助于我们了解影响基地环境的特殊因素，也有助于我们在设计过程中将它们考虑进去。

关于景观及
城市规划

对于室外空间的设计而言，最为重要的一点就是要利用地形来营造景观。无论什么样的地形，平的、坡的、台地的、层叠起伏的等，都时常会在空间设计和构建室内外联系方面给我们带来启发。如果一块用地有着一览周边景观的开阔视野，那么究竟是在用地本身上做文章，还是在景观上做文章，或是在现有周边景观中找关系，究竟哪种方案会更有趣，这就需要好好研究一下了。见图5 在以人类活动为主的环境中，人文因素对设计的影响与自然因素所带来的影响是同样重要的。许多建筑、街道和树木都给室外空间的设计提供了启发。

关于历史

对场地特性的关注不要仅限于那些直观的空间要素。无论采用什么方案都要在构思未来的同时对用地的历史作出回应。对于现有环境的塑造和改变是一种干预行为，并且会持续地对环境施加影响。但是我们应该牢牢记住，这种关联关系总是与建设项目的重要性相匹配的。

提示：
有关如何在背景环境中寻找灵感的详细介绍，可以参阅由 Bert Bielefeld 和 Sebastian El khouli 合著的《设计概念》一书，本书由中国建筑工业出版社于 2011 年出版（征订号：20273）。

提示：
在建筑与室外空间的二次规划中，很重要的一点就是，建筑师、城市规划师和景观设计师要密切协作，共同找到一个可以兼顾三方的设计方案（见图6及图7）

图 5：
果园成为景观特色

图 6：
城市的建筑天际线界定了城市空间

图 7：
建筑造型与植物形态之间的相互作用

图 8：
历史性场所里的现代形式语汇

也就是说，在建设诸如公园或者纪念馆这类具有重大社会影响的项目时，强调设计与历史事件之间的联系才会显得比较适宜和恰当。>见图 8

P20

植物的功能

植物具有诸多特性，正因如此，它们对于环境和人类而言就具备了诸多功能和意义。>见图 9 对人类而言，最为重要的一点就是在通常情况下，植物能否以及怎样才能在实现经济实用功能的同时还能获得令人满意的外观效果（如形态、色彩等）。对我们人类而言，植物的

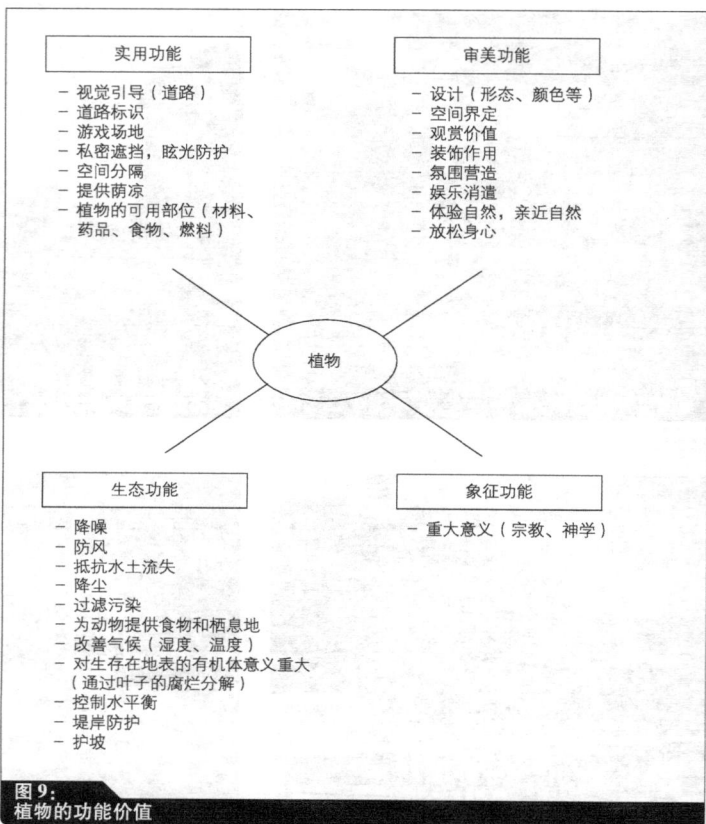

实用功能	审美功能
– 视觉引导（道路）	– 设计（形态、颜色等）
– 道路标识	– 空间界定
– 游戏场地	– 观赏价值
– 私密遮挡，眩光防护	– 装饰作用
– 空间分隔	– 氛围营造
– 提供荫凉	– 娱乐消遣
– 植物的可用部位（材料、	– 体验自然，亲近自然
药品、食物、燃料）	– 放松身心

植物

生态功能	象征功能
– 降噪	– 重大意义（宗教、神学）
– 防风	
– 抵抗水土流失	
– 降尘	
– 过滤污染	
– 为动物提供食物和栖息地	
– 改善气候（湿度、温度）	
– 对生存在地表的有机体意义重大	
（通过叶子的腐烂分解）	
– 控制水平衡	
– 堤岸防护	
– 护坡	

图9：
植物的功能价值

外观（特征、叶片、花朵、果实）具有极高的体验价值，不要低估它们对于人们头脑和心灵的重要作用。植物在保护生态和调节气候方面也发挥着根本性的重要作用。而审美功能、生态功能及实用功能之间并不存在相互矛盾的问题，而是可以共同发挥作用。

塑造空间的功能

室外空间主要是通过植物来塑造的。准确地说，植物搭建了丰富的层次并且在高度上形成了多种变化（从大树到球茎花卉）。无论是植物组群还是孤植的木本植物都能起到将不同的功能空间联系起来的作用。我们可以采用成组的植物或是一系列植物来塑造空间的尺度和形态。〉见空间结构一章

道路标识

植物具有标识道路的作用，它们可以用作路标、地标或是起到必要的提示作用（例如，用于提示斜坡的边缘）。〉见图10 对于小径或是

图 10：
一组树加上一个醒目的十字架就具备了道路标识的功能

道路而言，绿篱、独立的木本植物群或是较大规格的树都可以起到一定的视觉导引的作用。

防护作用

植物可以通过很多方式有效地发挥抵御恶劣天气及环境不利因素影响的作用（例如噪声或是大风等）。巨大的树冠可以在夏天让人们免受强光和酷热之苦。在冬季，掉光叶子的枝杈不仅不会遮挡人们的视线，也不会遮挡阳光。如有需要，密实的灌木绿篱可以完全起到减少或者降低风速、噪声和扬尘的作用。而那些生长在坡地或是堤岸地表的植物可以防止水土流失的发生。

植物通常都会同时具备多种功能。例如，无论是人工修剪的树篱还是自然生长的木本植物都可以用来划分停车场内的停车位。当利用它们来划分成排的停车空间时，它们既限定了空间又起到了防护作用。巨大的树冠在停车位上方形成了伞盖，可以在夏季起到遮阳的作用。

氛围与娱乐

当我们在着手自家花园的设计时，绝大多数人只会对设计本身感兴趣。而庭院的设计应该是既能实现预期的功能，又能兼顾美观的设计。通过精心设计可以让庭院营造出一些特定的感受，从而使其产生某种特别的效果和气氛。这种感受可以是宁静感，也可以是闲适感、放松感、安全感或是归隐感。人们可以从这些经过精心设计的植物中获得身心的愉悦，在充满和谐和惊喜的场所之中找到幸福感。花园和公园在形式上都是我们这个世界的理想化缩影。它们激发了人们的想

图 11：
成组的大树不仅营造了清静之所，还营造出很好的环境氛围

图 12：
大片的草地提供了游戏和活动的场地

象，不断去尝试缩小与世界上最早的花园——或曰伊甸园——之间的差距。此外，花园和公园往往都会反映出所处时代的一些特征。因此当前的社会、设计、经济、生态以及功能等因素都应该给予适当体现。〉见图11和图12 截然不同的花园、城市空间和广场都可以采用相同或是相似的设计手法，其中差别就要看在植物设计中侧重去表现什么了，例如，城市的特点就在于它的多元化和共生性。正是出于对不同场地的多样性以及场地不同品质和情调的表现从而催生了城市景观建筑学的出现。〉见图13 植物的外观在其中扮演着重要的作用，它们可以

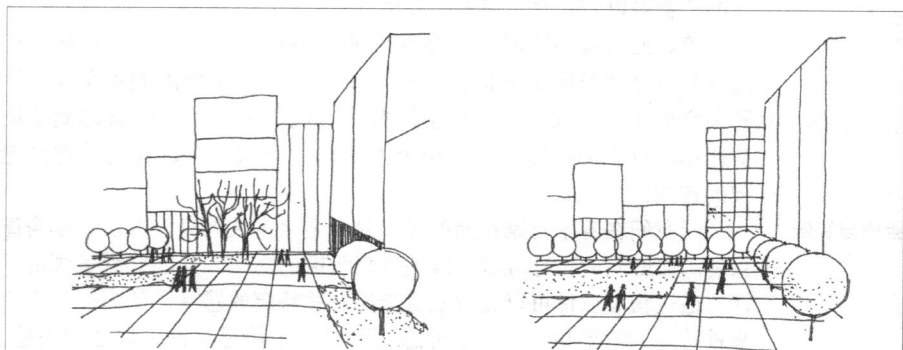

图 13：
植物的生长形态决定了空间的特色

106

表 2
植物的特性

明亮的	/	阴暗的
庄重的	/	随和的
安静的	/	喧闹的
苗壮的	/	瘦弱的
严整的	/	轻松的
分散的	/	集中的
规矩的	/	自然的
丰富的	/	单调的
天然的	/	人工的
开阔的	/	纷杂的
粗犷的	/	细腻的
雄伟的	/	精巧的

让广场、花园或是公园具有多种不同的特质。〉见表 2 例如，自然生长的树木会让整个环境显得自然而优美，修剪整齐的树木则会给人一种规整庄重的印象，而多变的颜色、质地和形态可以创造出丰富的整体形象和多种不同的氛围。〉见"植物也是一种材料，植物：外观"一章

在选择适用植物时，场地的位置、土壤的结构、气候条件以及预计在养护上投入的成本等因素都发挥着决定性作用。此外还要务必谨记，对于任何一个花园而言，不论其规模大小，都要给予适当的控制和关照，以便在多年之后，花园的形态也不会太过走样。不要让植物生长得过快过密，这会让花园显得杂乱无章，也不要在种植上随心所欲不讲章法。无论设计采用了什么样的风格，多年之后，人们应该依旧能够从花园中看出"园丁"（gardener）的个人印记。这是花园设计应该遵循的原则，也只有这样，花园才能够展现出它特有的氛围和幽静气质。

方向与导引
　　道路设计的特点就在于不仅要能看见目的地而且还要能直达目的地。道路是最天然的引导方式，对于人类的行为而言也是最自然的引导方式。途经空间通过整合设计越是显得自然优雅，道路也就会越有趣味。路人也会因此在"直觉驱使"的作用下决定由此前往选定的目的地。但是，目的地也不宜过早地被人们看到，以免打消人们的热情而去选择抄近道走过去。植物、座凳和景点都可以起到标识的作用，它们可以用作路标、地标或是空间的标识（例如用于

图14:
带状绿篱形成了透视效果并指示了方向

图15:
树木的形态突出了它们的路标作用

划定草地的边界）。对于小径和道路而言，灌木、独树、木本植物群以及大规格的树都可以起到视觉引导的作用。〉见图14 尤其是成排的树在远处就可以指示出方向。〉见"空间结构，组群"一章

在设计曲线导引道路时，一定要注意不要为了设计曲线而设计曲线。每一条曲线的形成都应该是从实际现状地形或是景观环境（如地形、植物、优美的风景等）中得到的结果。〉见图15

P25

空间结构

室外场地的设计与建筑设计一样，它们的关注点都在于空间的塑造。人类需要这样的场所也寻求这样的场所，因为这类场所会让人们觉得自在，而且还能在某些方面给人们以安全感。室外环境中可用来划分空间的实体元素有很多，诸如地形、植物、构筑物等，它们在塑造空间方面往往也会起到一定的作用。即便是非常细微的处理也能营造出空间感，例如一条水沟或是一丛灌木，一块凹地或是一个低垂的树冠等。在新地块的设计中，当我们想要在周边元素构成的环境中打造出新的空间形态和空间特色时，去研究一下这片区域的历史有可能会让我们从中得到一些有益的启发。一个空间的平面可以是几何形的，但也并非一定如此。一个空间的设计越是能够摆脱周边环境、地域和功能的局限，那么在空间形态的选择上也就会越灵活。在构思空

间形态时，通常会设置一些对比关系，诸如宽窄对比、远近对比等。运用这种处理手法的目的就是要让人们产生常规与夸张或是封闭与开放的感受。

空间限定

景观中的空间是利用竖向的层次来界定的，也就是水平层次。用于标示边界的竖向元素一定要足够多。〉见图16 在城市中，空间是由树和建筑围合成的。当我们利用一排树将两栋建筑联系起来的时候，这两种元素就共同构成了一条边界线。当四条这样的边界线围合出一片区域时，就会形成一个闭合的树的空间。当更多的建筑加入到这些边界线之中同时再给这个围合出来的空间赋以某种用途之时，那么就得到了一个广场，而这个广场又会具有更多的公共功能。它可以成为一个交往的场所，也可以容纳一个市场。此外，平坦开阔的场地中的下沉区域也可以形成空间，在斜坡地上切割出台地的做法也能形成空间。在室外场地的设计中，通常仅需要对边界元素和水平元素进行简单的处理就足够了。〉见图17 如果想要让空间形态具有较高的可识别性，那么非常重要的一点就是要突出场地的空间轮廓。〉见图18 如果想要获得鲜明的空间形态，那么空间轮廓线的转折和曲线部位就必须清晰可见；〉见图19 如果想要让人们在空间中逗留较长的一段时间，那么就需要对这些空间进行相应的设计并配置相应的设施。哪怕只是一棵树或是一个爬满藤蔓的花架也能起到这个作用，只要能够把它们利用起来，它们就会逐渐具有凉亭或是凉棚的功能。室外空间几乎总是能够吸引人们的注意，因为在通常情况下，其目的就是为了营造景观。只需利用少数几种元素、植物和辅助物就可以形成空间的边界和轮廓。在细节处理上均衡协调的设计能够形成独特的气氛。〉见"基本设计原则，功能"一章 在单一功能的空间中，成组的大树或是单株木本植物都可以用作分隔元素或是连接元素。〉见图20 同样，空间的闭合边界也会因不同形态元素的出现而被打开。对于小路、街道或是小巷而言，独立成组的木本植物群或是稍大规格的树木都可以在空间营造中起到引导视线的作用。

图 16：
采用柱形树来界定空间

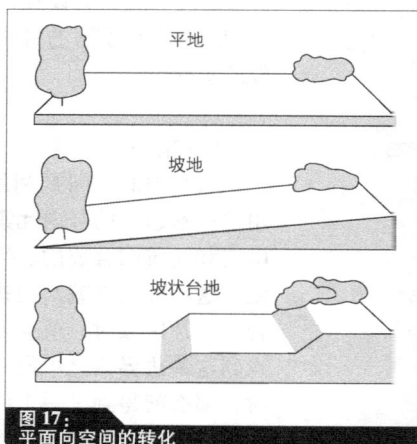

平地

坡地

坡状台地

图 17：
平面向空间的转化

图 18：
采用弧线型绿篱来突出空间的轮廓

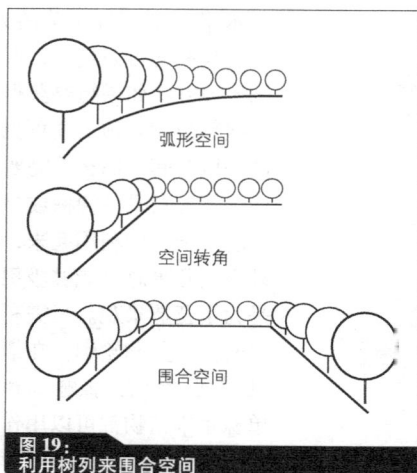

弧形空间

空间转角

围合空间

图 19：
利用树列来围合空间

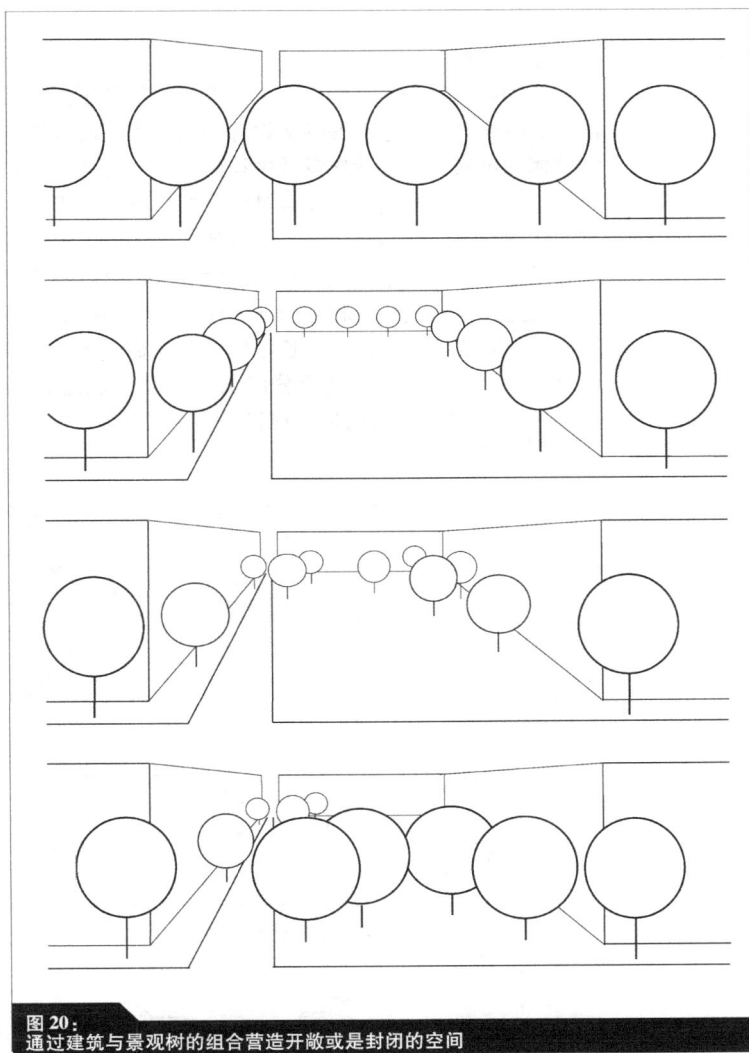

图20：
通过建筑与景观树的组合营造开敞或是封闭的空间

空间组织

植物设计就是植物的组织。只有当空间和平面的结构形态都非常清晰的时候，人们才能够领会设计师的意图，而室外空间的功能也才能够体现出来。但是，在设计中构建秩序的目的并不是要得到千篇一律的结果，因为同质化的元素是没有办法进行组织的。而对于种类丰富的植物而言，就需要建立一个清晰的秩序了。植物设计最为重要的一点就是要正确地去组织植物。这就要求在设计中不仅不要让植物的效果彼此产生冲突或是相互削弱，还要让它们能够互相突出对方的效果。这就需要依据植物的生长习性以及植物形态与色彩的设计搭配等要求，将每一种植物都进行分门别类。）见 "植物也是一种材料，植物：外观" 一章竖向空间的组织是种植设计中非常重要的一项内容。贴近地

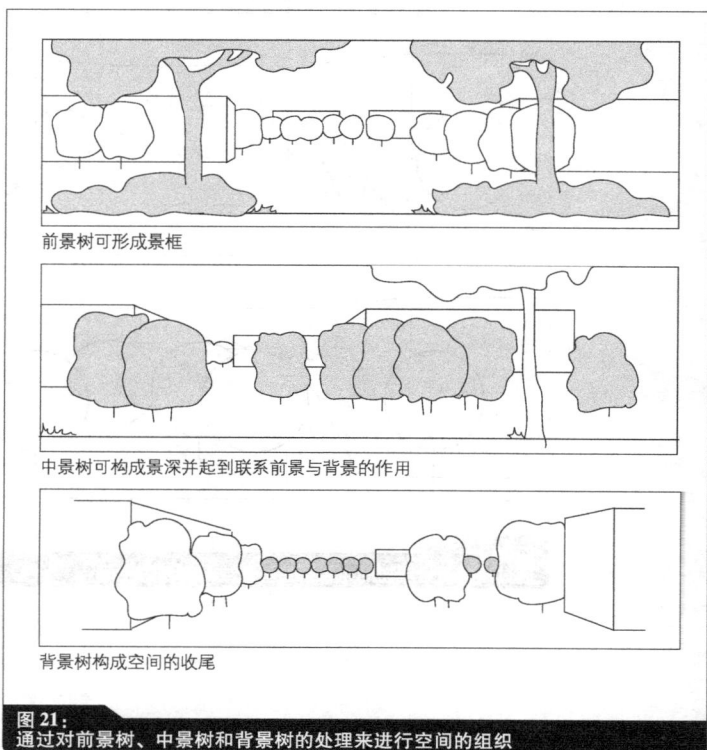

前景树可形成景框

中景树可构成景深并起到联系前景与背景的作用

背景树构成空间的收尾

图21：
通过对前景树、中景树和背景树的处理来进行空间的组织

图 22：
利用低矮的土丘和单植树来划分和分隔空间

图 23：
利用场地标高的变化和不同特点的植物来组织空间

表的低矮绿植与高大树木的组合总是能够形成视觉吸引力。需要注意的是，不同的植物可能会共同生存很长的一段时间，因此一定要熟记不同植物对光照的不同需求以及植物会随着生长在高度上发生变化这一现实情况。所以要尽可能避免植物之间出现生存竞争的问题。

前景，
中景，
背景

对于一个看上去结构清晰的室外空间而言，其空间的组织就体现在有意识地利用植物营造出了前景、中景和背景层次，并且很好地处理了观赏距离的变化对植物规格与色彩之间关系的影响等问题。前景树和背景树所发挥的作用是完全不同的。前景树就如同建筑一样。它们给人们提供了阴凉，可以让人们在这里悠闲地欣赏风景。中景树形成了比例关系和空间景深。背景树构成了围合空间的界线。背景绿化承担着将整个花园空间创造性地统一起来的重要作用。要想突出空间效果，形态复杂的前景绿化，如花坛或是草本植物区等，需要搭配一个安静的空间背景。背景绿化通常都具有两个功能：其一是底衬功能；另一个功能则是将周边环境和左邻右舍整合起来并在视觉连成一片。〉见图 21

将空间范围内多种不同功能和设计手法（例如，主题公园）的区域关联起来（空间序列）也是空间组织的一种表现。〉见图 22 其中有一点需要特别注意：由植物形成的室外空间与建筑空间的不同之处就在于构成尺度关系的元素会因为生长而发生变化。通过定期修剪可以很好地维持空间的尺度，这对于那些具有建筑气质的花园而言尤为必要。〉见图 23

边界

围合空间的手法多种多样。建筑、墙体、围栏、绿篱甚至是地面形态都能够形成统一的实体边界。复合边界则是采用多种元素沿边界线成排布置而形成的，这些元素可以是一棵棵单植的大树、一株株单植的灌木、爬满攀缘植物的构筑物、特殊地形、石头、影壁墙、小土丘等。当使用植物来围合空间时，就会遇到前面提及的一种情况：植物会随着时间的推移而发生变化。它们会不断生长。若是置之不理，它们的生长形态就会发生改变。也就是说，如果一座花园在种植完成后的头几年看起来仍然宽广开阔的话，那么二、三十年之后，这里就会变成一个令人十分不适的幽闭之所，这通常都是因为缺少维护和修剪的缘故。因此，当我们在设计中需要选择植物或是为植物设计生长环境的时候，就很有必要知晓所选植物的最大生长高度、可能发生的生长行为以及生长环境可能带来的影响（例如，是孤植还是群植）等情况。每年的定期养护，例如给木本植物疏枝等措施，不仅可以避免自然生长的绿篱和灌木过度生长，还可以减少植物底部生长不良的问题。边界的高度及围合出的空间尺度决定了人们能够看到的天空范围，我们可以借此来营造开阔或是狭小的空间感受。一道两米高的绿篱会随着观赏距离的增加而逐渐丧失空间感，这就意味着，随着围合空间的尺度的增加，围合边界的高度也需要相应增加。用来塑造空间的地形（如沟渠、地垄、台地等）在这层关系上表现得尤为突出。用来界定空间的元素不论是高出人们的视线还是低于人们的视线（大约 1.7 米）都是很有必要的。〉见图 24 当低矮的围合元素，如绿篱、花棚、踏步、路缘石等与人们的视线发生交叉时，它们也会对室外空间的塑造发挥一定的作用。它们完全可以用来指示花园的边界或是用来划分不同功能的场地。与此同时，它们还可以很好地与周边环境建立联系，因此，使用这样的元素所塑造的花园空间会在视觉上得到延展。另一方面，经过修剪的树墙或是自由生长的树列在构建空间骨架的同时也对空间做出了清晰的划分。可以用来限定空间的植物形态以及配有攀缘植物的空间限定元素多种多样，这就要看我们想要得到什么样的质感效果和通透效果了。可供选择的元素从利用生长不明显的植物卷须构成的通透精巧的格架到用绿篱修剪而成的巨型植物墙以及种植墙等。〉见图 25 及 26 悬垂植物、丛生的草本植物以及侧枝垂到地面的大树都能够阻挡人们的视线，而高大的树木或是分枝点高的大型灌木就可以让人们的视线自由穿行。到了冬天，则会形成全新的空间效

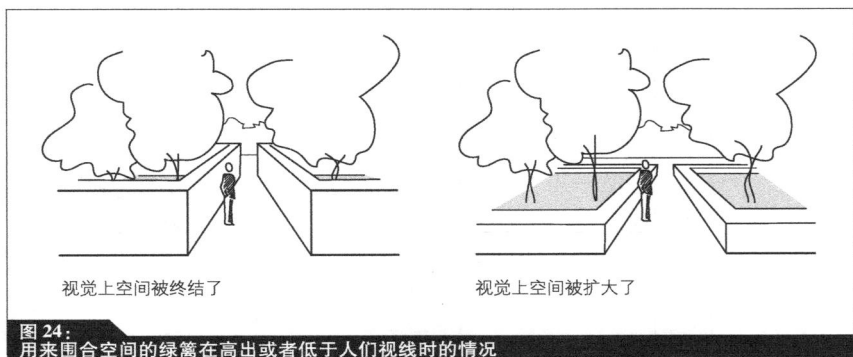

视觉上空间被终结了　　　　　　　视觉上空间被扩大了

图24：
用来围合空间的绿篱在高出或者低于人们视线时的情况

图25：
修剪整齐的绿篱形成了一个空间轮廓

图26：
爬满葡萄藤的廊架——构筑物作为空间的边界

果，大树的叶子在这时都已掉光，灌木则开始成为主角。从植物的明暗对比以及色彩与质感的对比效果出发，选出的植物要能够与设计场地（诸如游戏场、庄重的建筑、墓地等）的整体风格相匹配。〉见"植物也是一种材料，外观"一章

P31　　　　**组群**

要想在彼此分开的视觉元素之间建立起联系，可以利用它们的共同点来形成队列或者组群。例如，成组布置的树就可以形成空间上的联系。树的布局可以采用多种不同的手法，既可以让它们构成规矩严整的树阵，也可以采用放松自由的手法让它们形成轻快的树林。

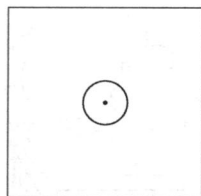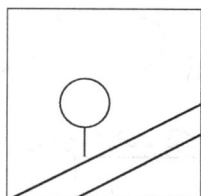

独树

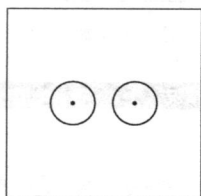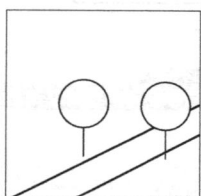

双树

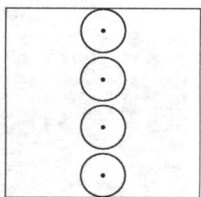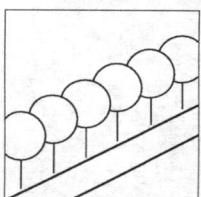

单排树

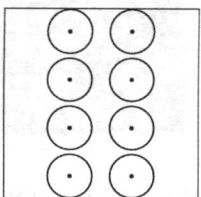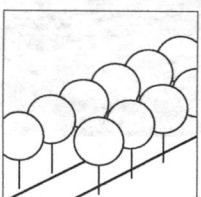

双排树

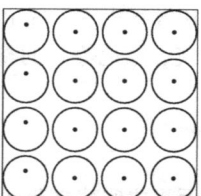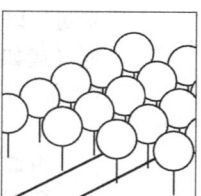

树阵

图 27:
树的规整组织形式

图28：
独树

图29：
双树

规整的组树　　　　规则整齐的组树在城市的室外空间中能够形成非常有冲击力的形象。当利用十几棵树共同组成一个图案时，我们称之为树阵而不是组树。这是一种非常简洁但又非常有效的设计手法。〉见图27

独树　　　　　　成年的孤植大树在景观中的效果非常突出。它们是在远处就能看见的地面标志物。在花园的设计中，孤植的大树不仅可以形象鲜明地融入整体环境中，据守着重要的位置——道路的尽头或是视线的终点，作为视线的焦点或是构成花园空间的转角标识——还可以为了取得对比效果，有意将其布置在整个组织体系之外的区域。〉见图28

双树　　　　　　成对的大树在景观、花园和城市空间中也是一种设计元素。〉见图29在花园、出入口、乡村宅邸、花园洋房或是花园与花园之间的过渡区域中，通常都会采用成对的大树种在两侧起到强调的作用。

树列　　　　　　在许多欧洲的人文景观中，树是场地中最为重要的造型要素之一。树列也是园艺学中反复使用的设计元素。它们不仅可以界定空间，还可以突出韵律感。〉见图30在许多城市的河道、街道以及广场周边都种有呈线性排布的树列。它们比建筑界面的造型能力更为突出。因此，在塑造城市和谐的整体形象方面，利用相同的手法来设计绿植是一种非常重要的方法。树列在设计中的作用包括：

　　　　　　　—指示方向；

　　　　　　　—遮挡视线；

　　　　　　　—形成围合空间和线性空间；

图 30：
树列

—协调沿街立面。

　　如果建筑之间的差异很大或是街道的整体形象杂乱无章的话，那就可以采用树来充当视觉调节因素。另一方面，树也可以让看起来单调乏味的街道活跃起来。〉见图 31

林荫道

　　多排树可以形成林荫道。在与树相关的设计元素中，林荫道是让人印象最为深刻的元素之一。林荫道在德语中被称为 *Allee*，它源自法语的 *aller*（"去"），是指两侧都长有树的小路。〉图 32 在城镇中，人们可以在树下散步或是嬉戏。在道路两侧及中央绿化隔离带上种树的做法就意味着大部分的路面都是在树列下穿过。一些宽阔的林荫大道，如位于柏林中部的菩提树大道（unter den linden）等更是举世闻名。这样的道路绿化可以让城市和每条街道的形象看起来都更加优雅，也更有艺术气质。林荫道的文化史起始于文艺复兴时期，在 18 世纪达到顶峰。在专制主义时期，绵延几公里的僵直林荫道成为人类主宰环境的一种体现。拿破仑大街（Route Napoleon）的杨树林荫道就是这样的一个例子。君主和地方统治者们把道路修到了自己的城堡、乡村宅邸和狩猎小屋前，道路的两侧种满了遮阴的大树。对于道路种植而言，树的间距建议控制在 5 ~ 15 米，具体距离要看采用什么树种。树的间距越近，空间的围合感就会越强。

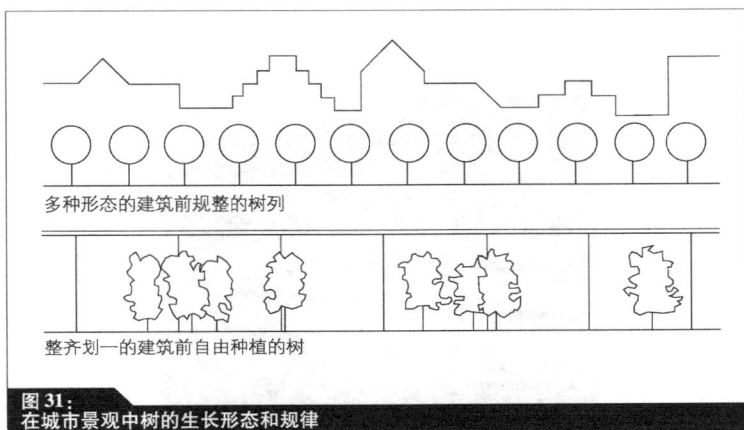

图 31:
在城市景观中树的生长形态和规律

多种形态的建筑前规整的树列

整齐划一的建筑前自由种植的树

图 32:
林荫道

树阵

      树阵是采用相同树龄和品种的树种在一起所形成的。它们在各个方向上的间距都是相同的，通常它们都是种在场地中的一块方形区域里。如果采用树冠浓密的落叶树种，如七叶树和枫树，就可以突出这种规则种植的建筑感。如果采用修剪成立方体的椴树，那么它们呈立方体形的树冠就可以让树阵呈现出一种近乎于"建造出来"的感觉。在城市环境中，树阵成为建筑学的组成部分。朝同一方向延伸的多条树阵带可以形成强烈的轴线（如林荫道）。当规整的树阵出现在生长形态相对自由的木本植物群中时（如景观公园），就会给人们一种建筑意味（绿树环绕的建筑物或几何形广场）的联想。

图33：
树阵广场

树阵广场

　　当十多棵树按照同一种规律进行种植时，它们就不再是普通的组树了，而是形成了一个树阵。〉见图33 不过，组树在城市中的作用并不仅仅是构建空间的元素。它们还有一些其他的重要功能，诸如：

　　—聚会广场；

　　—树荫广场；

　　—提供广告场地的广场；

　　—可以改善局部微气候的广场。

　　树阵广场非常受人们的喜爱，因为来这里的人们可以选择停留在阳光下或是树阴里。特别是在夏天的那几个月里，树阴在午后和傍晚都会让人们心存感激。还可以采用喷泉、五颜六色的花圃、绿篱、灌木、片墙在广场上组成各种图案来提升广场的品质。

树的造景组合

　　以植物为主题的城市设计具有双重审美功能。它不仅可以改变城市的面貌，而且也是自然融入城市的一种体现，城市也就因此不再是自然中的一块"补丁"了。〉见图34 这类设计最重要的作用就是突出了与人工形态的差异，如建筑物。几个世纪以来，城镇都被建设成包括网格形以及严整的规则形态在内的几何图形。在这个刻板的结构体系中，成组的景观树代表了自然的一角，它们与理性的城市建设法则形成了直接的对比。树的造景组合方式有很多种：

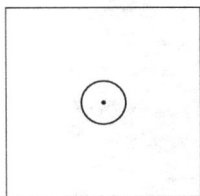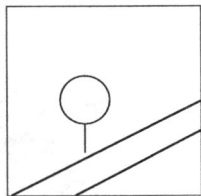

单株树

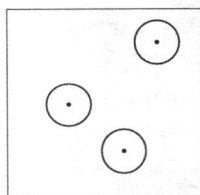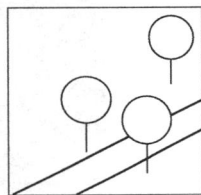

多棵单株树

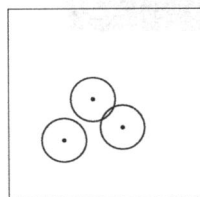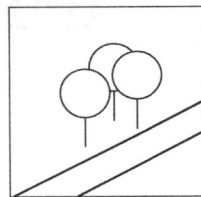

三棵树的组合

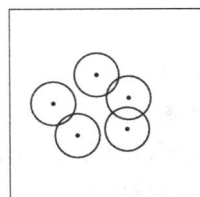

五棵树的组合

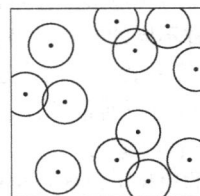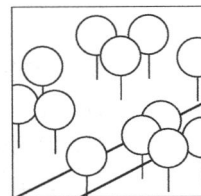

不规则树丛

图34：
树的造景组合

121

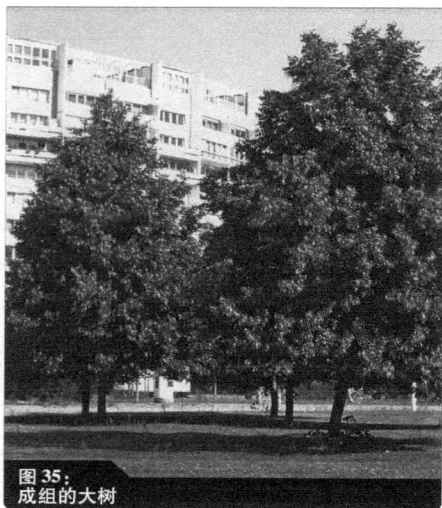

图35：
成组的大树

—单株树；

—多棵单株树的组合；

—三棵树组合；

—五棵树组合；

—树林。

组树

组树与林荫道中的树列在功能上有一个不同之处。它们不仅能够突出建筑物，还能够围合出小面积的区域，或是在城镇的总体格局中成为补充过渡的部分。设计师可以利用自由形态的组树来引导人们的视线或是通过恰当的布局设计来增加室外空间的景深。>见图35 种植得当的组树和地形之间的相互配合可以形成非常优美的花园景观。纵览园艺学几个世纪的发展和各种不同的园林风格，我们会发现组树这种设计元素的使用频率很高，其中既有自由形态的，也有规则几何形的。在几何形态的组树中，树间距通常都比较近，在 1.5～2.5 米左右，因此也被称为树团（tree packages）。

少量的几棵树或是松散的组树可以界定单栋建筑或是在多栋建筑之间建立起视觉联系。>见图36 树可以营造出空间景深，还能弱化建筑的生硬边缘。少量的几棵树或者一小片组树可以在视觉上让不同形态

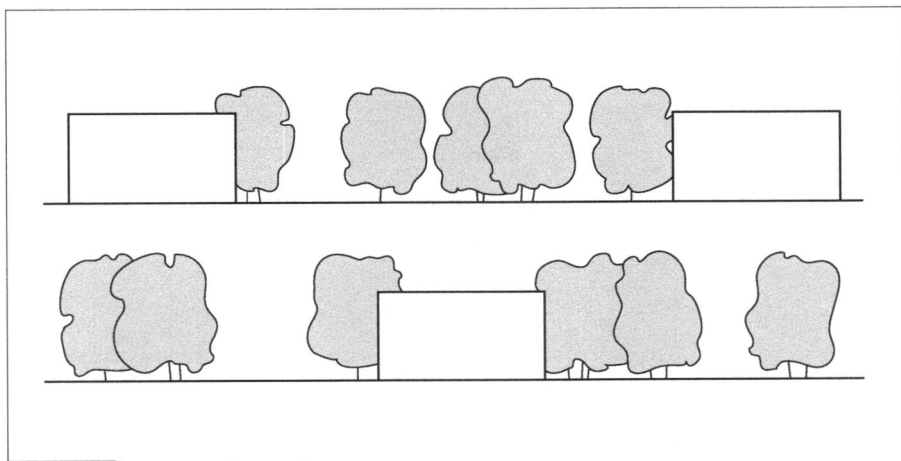

图 36：
设计优美的组树可以在多个建筑之间建立起视觉联系或是用来框定单个建筑

的多栋建筑取得协调。〉见图37 当建筑造型众多或是在房地产开发项目中，只需少量的几棵相同品种的树就可以给人们留下秩序井然的印象。这就很好地说明了树木是如何让居民和到访者对一个城市街区产生注意并形成认知的。

树林   树林的特点多种多样，还可以营造出多种氛围。这就要看我们去选择什么种类的树（例如，是选择透光性好的还是遮阴的，叶片是深绿的还是浅绿的，叶片是有光泽的还是无光泽的等），此外还要看树的间距和种植形态（严整的、规矩的、灵活的、不规则的）。树林是采用相同品种和树龄的木本植物组合而成的。采用树冠松散、质感细密的树种（如桦树、落叶松、松树、刺槐）或是搭配低矮的地被植物、草地或是草坪都可以突出树林的开放感。〉见图38 不同的树种、不同的树龄以及不同的树间距会形成完全不同的气氛。例如，由高大的山毛榉组成的浅色树林看起来会觉得明亮而亲切，而有着常绿的针状外观的松林看起来就显得相对阴暗。与严整规则的树林相比，种植灵活自然的树林不会受到任何规则格网的限制。树与树之间的距离控制依循"分散种植"的原则。光照区与树阴区的交替关系也是没有规律的。有的地方宽，有的地方窄，疏密交替。自由形态的树林可以营造出多种感受，这取决于选用什么样的树种，营造出的感受可能是田园牧歌型的，也有可能是忧郁阴暗型的。

123

图 37:
利用树来统一建筑的不同风格

标高/高度的设计

　　在室外，高度上的差异具有很强的空间限定作用。高度上的变化——不同高度间的过渡——可以形成空间的边界。高度的变化既可以是界面清晰的变化，也可以是柔和的过渡（造型）。在室外场地中，空间的高度通常都是利用植物来构建和形成的，是通过不同高度

图38：
种在起伏地面上的明亮的桦树林

图39：
作为空间底衬的草地

的植物（从草地到树木）之间的分层和分界来实现的。草地、草坪、地被植物以及草本植物给人们的印象是平坦的，强调的是平面的标高。而灌木、绿篱、大型灌木、单株木本植物和大树才是限定空间的主角，这是因为它们是有高度的。而攀缘植物如常青藤、弗吉尼亚爬山虎等，可以在墙面和建筑上形成大面积的绿色。

地坪高度　　　　修剪整齐的大片草坪、草地、同种草本植物组成的低矮的地被层，以及木本植物、膝盖高度的绿植等，由于它们的质感、色彩和结构都不相同，因此它们的水平表面也有着不同的视觉效果。它们可以营造出多种气氛，庄重的、平和的或是漫不经心的等。对花园而言，草坪就是空间的基底。〉见图39 一个修剪整齐的草坪会因其柔软纯净的质感让人们感觉酷似一块平铺的地毯，它让室外空间看起来很宁静。因此，草坪要尽可能的开阔和连续。修剪很短的草坪则会让地形的变化一目了然。此外，草坪还会在一年之中不断变换着它们的外观。草地中占比最大的草和花的颜色会随着季节的变化让草地呈现出各种不同的视觉效果。而随风起伏的草浪还会让这种效果进一步得到突出。草坪的修剪也可以成为一种设计手段，例如，在草坪上修剪出小路、仅修剪草坪的部分区域或是仅修剪草坪的周边区域等方式来形成有趣的对比效果。〉见"植物也是一种材料，设计原则"一章以及"植物也是一种材料，植物：外观"一章

由常绿植物、低矮的草本植物和木本植物所构成的地面植被的外观形态要比单纯的草坪更加生动，这主要是因为前者有着丰富的色彩

125

图40：
树和绿篱可以用作限定空间的元素

和质感。植物的叶片越小、植株越矮，就会越发显得平整。

限定空间高度的元素有草坪、草本植物、小灌木、绿篱、大灌木、单株木本植物、组树和绿植墙。在花园或是园林的草坪上种植灌木和绿篱可以起到破解和划分地面的作用，并能因此营造出空间景深。〉见"空间结构，空间限定"一章如果需要营造一道视觉屏障，可以考虑采用加大灌木绿篱之间间距的方式来形成空间的背景。〉见"空间结构，空间组织"一章灌木可以在大树与靠近地面生长的植物之间形成过渡，还可以在开阔的景观与公园或是绿地之间形成过渡。灌木最为重要的一个特点就在于它们的叶片。如果灌木能在人们的视线高度上铺开很大一片区域，那么对于这块绿地给人们留下的整体形象而言，灌木的叶子发挥了巨大的作用。这种作用可以是低调的，也可以是醒目的。〉图40 在一年之中的大部分时间里，灌木都会非常引人注意（如叶片的颜色、花、果实等），但是它们的形态通常都不是很突出。

除了树之外，修剪整齐的绿篱也是一种非常重要的设计元素，因为它们把规整的元素带入到设计和形态空间中，而且其轮廓、造型和质感效果都非常突出。〉见"空间结构，边界"一章以及"植物也是一种材料，植物：外观"一章修剪绿篱有以下几种外观形态：

—绿篱空间；

—连续绿篱；

—绿篱挡墙；

—块状绿篱；

—自由形态的绿篱。

窍门：

　在住宅旁边种有大树和灌木的绿地里种满同一品种的植物时，例如地被类植物，不仅可以将不同的地块统一起来，还能让种植设计显得很纯净。

　　绿篱空间是利用高出视线的绿篱植物围合而成的。单独一种植物就可以围合出空间。连续的绿篱可以长到半人高，还可以布置成扇形、弧形以及其他一些奇特的图形。绿篱挡墙可以在室外场地中用作绿色的边界或是"影壁墙"。它们可以种在远离建筑的地方或是与建筑形成对比，具体处理手法并非是一成不变的。绿篱块是由立方体形的绿植组成的不同高度的造型。多排种植的方式会让绿篱的景深感更强烈。修剪成自由形态的绿篱会产生强烈的雕塑效果。在比利时景观建筑师雅克·魏茨（Jacques Wirtz）的设计中，经过修剪的绿篱是用来表现风格的元素。

垂直高度　　　　垂直的表面例如建筑墙面或是单片墙体，以及藤架、凉廊、影壁墙这样的垂直构件都能形成封闭或是通透的空间边界。它们的表面可以局部或是全部爬满攀缘植物，并因此变成绿色的或是开满鲜花的空间墙体。当攀缘木本植物爬满整个建筑外墙或是单独一片墙体的表面时，它们就会形成一种非常有趣的肌理效果。绿色的表皮看似一件绿色的外衣。〉见图41 另一方面，局部种植攀缘植物可以起到突出特定区域的作用。花架、棚架和藤架会在植物的拥绕下产生柔和的过渡和诸多令人愉悦的细节。而攀缘植物则会让建筑呈现出一种独特的外观。木本攀缘植物可以爬满垂直构件而且只需占用极少的空间。依据攀缘方式的不同可将攀缘植物进行分类。自攀缘木本植物无需任何攀附物就能爬满垂直表面和垂直构件（也包括水平

图41：
墙面绿化形成了一个空间轮廓线

的）。而缠绕型和藤蔓型植物则需要一些支撑物来辅助它们攀
爬。〉见图42及表3

生长形态		攀附物		
自攀缘植物		墙 树 表面（水平面、倾斜面、 垂直面）		墙
			吸盘	攀缘根
缠绕植物		格架 棚架 金属网 水平或垂直张拉的线绳		
藤本植物		张拉的线绳 藤架 廊架		
漫生植物		墙 树		墙

图42：
攀缘植物的生长形态及其攀附支撑

128

生长形态	学名	名称	大面积覆盖	局部覆盖	生长高度	生长速度*	常绿	落叶
自攀缘型	*Hedera helix*	常春藤	×	×	10～20	s	×	
	Hydrangea petiolaris	藤绣球		×	8～12	m		×
	Parthenocissus quinquefolia "Engelmannii"	恩氏弗吉尼亚爬山虎	×		15～18	f		×
	Parthenocissus tricuspipidatra "Veitchii"	韦氏日本爬山虎	×		15～18	f		×
缠绕型（需要攀附物）	*Clematis montana*（及其变种）	铁线莲		×	5～8			×
	Clematis montana（Rubens）	鲁宾斯铁线莲		×	3～10			×
	Clematis tangutica	甘青铁线莲		×	4～6			×
	Clematis vitalba	葡萄叶铁线莲		×	10～12			×
	Parthenocissus quinquefolia	弗吉尼亚爬山虎	×		10～15	f		×
	Vitis coignetiae	紫葛葡萄		×	16～8	f		×
藤型（需要攀附物）	*Aristolochia macrophylla*	马兜铃		×	8～10	m		×
	Celastrus orbiculatus	东方南蛇藤		×	8～12	f		×
	Lonicera caprifolium	蔓生盘叶忍冬		×	2～5			×
	Lonicera heckrottii	金光忍冬		×	2～4			×
	Lonicera henryi	巴东忍冬		×	5～7		×	
	Lonicera tellmannia	金银花		×	4～6			×
	Polygonum aubertii	山荞麦	×	×	8～15	f		×
	Wisteria sinensis	中国紫藤		×	6～15	m		×
蔓生型	*Jasminum nudiflorum*	迎春		×	2～3			×
	Rosa by varieties	藤本月季		×	2～3	m		×

*s＝生长缓慢，m＝生长速度适中，f＝生长速度快。

当然，也可以使用苗圃中培育的绿篱植物来做出墙一样的垂直面。但是具体选择什么样的树种、希望让植物达到的最终高度、叶片的密度以及是否需要让绿篱在冬季或是夏季都保持绿色等，这些都是决定性的因素。绿篱每年至少需要修剪一次，以便维持其形态和密度。在远处可以用树墙形成边界线，以此获得视觉冲击力。

比例

比例与尺度的变化会影响空间的形态。"比例"是指不同物体之间在主要规格尺寸上的精确的（或是经计算得出的）关系（比如说高度和宽度之间的关系等），它是对设计元素之间的数值关系所进行的视觉衡量。通过改变比例可以让空间变浅或是加深，当然，利用透视也可以实现同样的效果。对于一条街道而言，如果建筑沿街立面的整体平均高度与行道树高度之间是3：5的关系，那么比例就会显得均衡协调。如果通过改变竖向尺寸让比例变成2：6或者4：4，那么就会产生全然不同的空间效果。在比例关系固定的情况下，可以利用树来调整建筑与人之间的尺度关系。因此，让树与建筑物的高度和体量取得协调一致是非常重要的。高大的树木适用于比较宽的道路，而且树与建筑之间的距离不要太近，而对于比较窄的道路而言，采用相对矮小的树木或是将树种在靠近建筑的位置效果会更好。〉见图43 如果尺度是呈逐渐递减的关系，就可以借助透视手法来强化空间的深度和宽度。通过增加能够在视觉上形成空间景深的插入元素、层次元素或是线性元素，就可以得到一个不错的空间景深效果。借助特殊的规格和特征，这些元素也能够对整个空间的比例施加影响。孤植的大树、花架以及反射天光的大面积水体等都可以达到这个目的。〉见"空间结构，空间组织"一章线性结构，例如木本植物绿篱或是由叶片和花簇形成的五颜六色的效果也可以营造出透视感。

场地的地势和地形对室外空间的比例会产生很大的影响。例如，从人们所站位置向斜下方延伸开去的场地会显得比较远，这是因为场地从人们的脚下向后越退越远。下坡的场地会显得比较开阔。相反，上坡场地会显得离人比较近，由于所有的东西都在你的视线范围内，因此在视觉上会觉得它正在向我们走过来。〉见图44

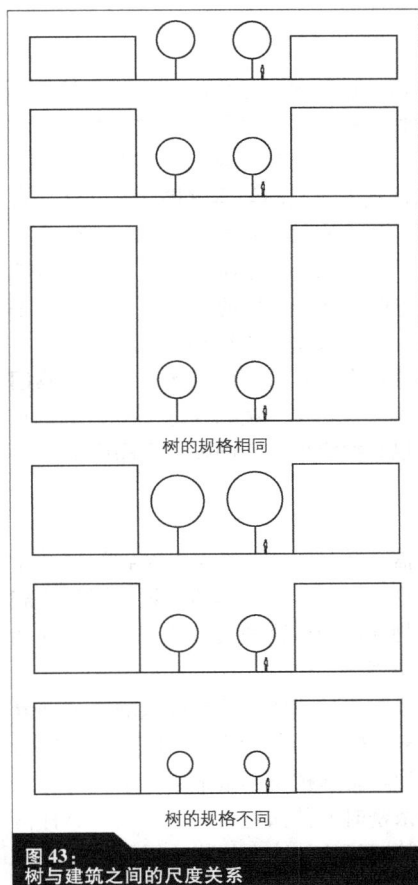

树的规格相同

树的规格不同

图 43：
树与建筑之间的尺度关系

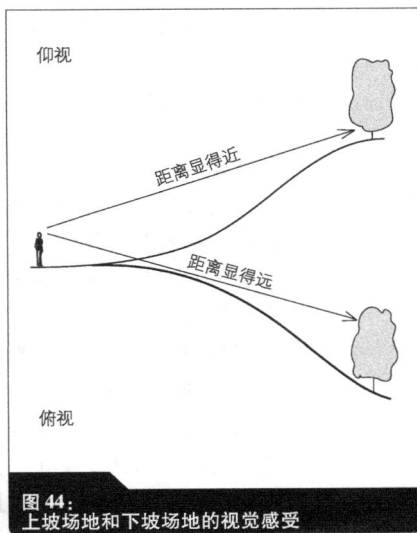

仰视

距离显得近

距离显得远

俯视

图 44：
上坡场地和下坡场地的视觉感受

✎

窍门：

　　红色和橙色能让物体显得靠前，因此它们具有缩短视距的效果；蓝色、蓝绿色及蓝紫色会让物体显得靠后，因此它们具有增加视距的效果（见"植物也是一种材料，植物：外观"一章）

✎

窍门：

　　公园、小镇广场和花园为研究植物及其应用提供了大量的机会。通过对植物的观察和分析可以认识多种多样的植物，了解它们的不同特性。某种植物的特点是什么？它是怎样与其周边环境和谐相处的？通过对好的实例和不好的实例（！）进行分析学习，等到我们自己着手做设计时，就已学到很多方法了。

植物也是一种材料

　　挑选适用的植物种类是一个需要进行多方考量的过程。非常重要的一点是，当把不同外观的植物种在一起时不仅要让植物之间相互协调，还要让植物与周边环境取得协调，并由此形成一个有着强烈视觉冲击力的鲜明形象。千万不要忽视室外景观的整体效果，这一点非常重要。要想做好植物设计，就要具备扎实的植物知识。例如，生命力旺盛的植物会迅速竞争掉左邻右舍并最终蔓延到整个花园。结节草（Knotgrass 蓼属）就具有这样的特点。有些植物的生长非常缓慢或是与其周边植物不能兼容生长，因此就会长得比较矮小。

　　植物是有生命的材料，通常都不会按照我们的设想来生长。我们的设计只是给出了一个基本的框架，让选定的植物在这个框架下生长发展，再通过定期的维护，就可以掌控室外空间的发展和品质了。

植物：外观

　　好的设计通常都很简洁，只需少数几种精心选出的植物类别及品种即可。这就要求我们对所选植物的外观有着准确的认识，同时还需要运用一些审美法则。在选择植物和确定植物种植位置时，运用这些原理知识要比完全依赖直觉可靠得多。

形态

　　植物的形态是由植物断面轮廓线所形成的。植物枝杈的末梢部分（如树梢、叶片、花朵）越是浓密，植物的外形及结构就会越鲜明。在夏天，落叶树及落叶灌木的外形与常绿树及小型常绿木本植物的外形同样醒目，当其表层叶片特别浓密时尤其如此。到了冬天，那种树枝和树梢生长得比较细密并且外轮廓线比较清晰的落叶树及落叶灌木的外形也会很醒目。植物的外形越是简单鲜明，就越是容易被人们所识别、描述、描绘和设计。植物的生长方式不同，其形态的分类也会有着显著的差异。〉见图45 形态既可以指整体也可以指表面。许多连续的造型都是从最简单的基本图形中衍生出来的：如方形、圆形、三角形等。而自由形态的造型就要明显复杂得多。如以一株成年雪松的枝干结构为例，人们可以从它发散式的细密的针状叶片上看出它的形态特点。

形态特征

　　一株外形处理得当的木本植物会因其具有建筑气质和艺术效果而引人注目，它们可以用来装点公园、花园或是绿地。还可以依据它们的生长形态和生长方向来塑造静态或是动态的形象。植物的形态特征

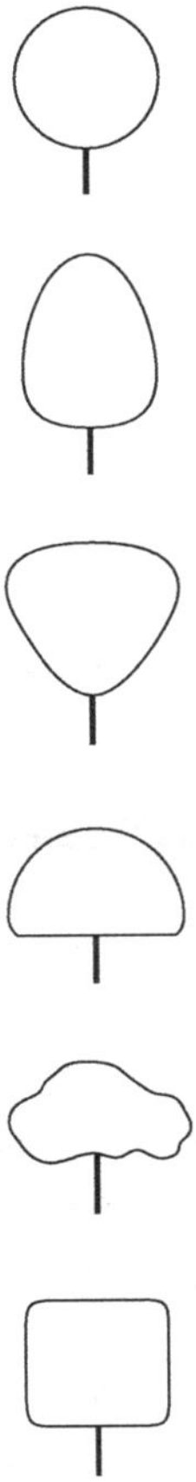

分类	示例	用途
球形	尖叶械 "Globosum" （挪威械）	紧凑的小型树，适合用在围合空间和前庭花园中
蛋形	心叶椴木叶 "Erecta" （小叶椴）	采用单排或双排的种植方式用于讲求秩序的场合，适用于城市开放空间
漏斗形	山樱 "Kanzan" （日本樱）	适合树列及树阵等种植形态
伞形	美国梓树 （黄叶美国木豆树）	成年大树适用于有遮荫要求的坐凳区及小块场地
松树形	欧洲黑杨 "Austriaca" （修剪规整的澳洲黑杨）	在开阔的丘陵以及山地景观环境中，其外形效果非常突出
方形	阔叶椴 （修剪规整的阔叶椴）	适合用于讲求秩序的场合，塑造带有建筑气质的植物造型

图45：
树的外形

可以分为"无方向型"、"方向固定型"以及"方向不持续型"。球形作为一种简单的设计图形是没有方向性的，并且有一种静态感。水平以及垂直生长的植物由于生长方向是固定的，因此也会让人们觉得是静态的。而攀缘植物及悬垂植物则属于生长方向不持续型，它们发散式的生长形态看起来会有一种动态感。无论静态的还是动态的效果都

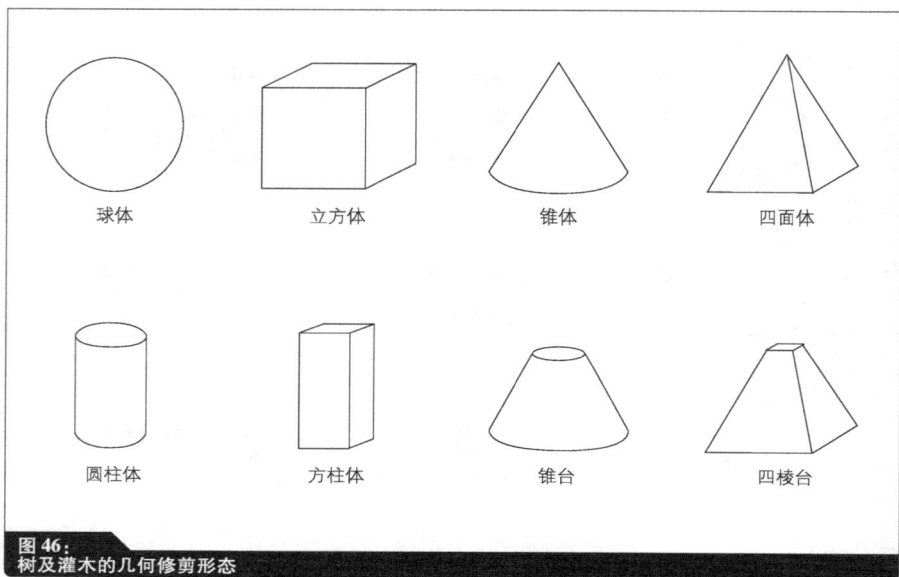

球体　　　　　　　立方体　　　　　　　锥体　　　　　　　四面体

圆柱体　　　　　　方柱体　　　　　　　锥台　　　　　　　四棱台

图46：
树及灌木的几何修剪形态

是可以通过形态的差异组合和对比组合来予以强化的。例如，当竖向形态（柱形）的植物沿曲线道路排布时（方向不持续）就会显得相对稳定。相对于高起的建筑（高层建筑）而言，水平展开的植物形态（树列）可以形成一个横向的对照，还可以利用球形植物（无方向）在建筑一侧组成一条鲜明的弧形植物带（方向不持续）。〉见"植物也是一种材料，设计原则"一章

修剪型木本
植物
　　整齐规则的修剪形态会使得树、灌木和绿篱的轮廓线变得完整清晰。我们可以把挑选出的落叶植物及针叶植物修剪成几何形（立方体、柱体、球体、四面体、锥体、锥台等）或是自由形。〉见图46 严格来说，修剪造型可以分为盒子形、盖子形、格架形以及球形。〉见图47 修剪后的绿篱可以形成连续、清晰的空间边界。低矮的修剪绿篱不仅可以围合出一小片花园，而且也不会阻挡人们的视线。规整的修剪方式不仅可以控制植物的体量，还可以让植物的体量几乎维持不变。修剪整齐的木本植物特别适合用在有建筑气质的花园景观中或是用来组织开阔场地的景观。〉图48 但是，规整的修剪方法只适用于非常有限的几种植物。〉见表4

 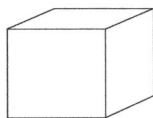  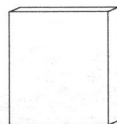

球形　　　　　　　　盒形　　　　　　　　盖形　　　　　　　　墙形

图 47：
树木的修剪形态

表 4：
适合进行修剪的树种及灌木

学名	名称	单株	绿篱	拱形	几何形	伞形	花架	盆景
落叶型乔灌木								
Carpinus betulus（及其变种）	欧洲鹅耳枥	×	×	×	×	×	×	×
Cornus mas	山茱萸		×			×	×	×
Crataegus（各类品种）	山楂	×						
Fagus sylvatica	欧洲山毛榉		×	×	×			
Platanus acerifolia	二球悬铃木	×						
Tilia（各类品种）	椴树	×	×	×				
常绿型乔灌木								
Buxus sempervirens arborescens	黄杨		×		×			
Ilex aquifolium（变种）	冬青				×			
Ilex crenata（变种）	日本冬青				×			×
Ligustrum vulgare "Atrovirens"	野生女贞		×		×			
Pinus（各类品种）	松					×		×
Taxus（各类品种）	紫杉		×		×	×		×
果实类乔木								
Malus domestica（变种）	苹果						×	
Pylus communis（变种）	梨						×	

图 48：
修剪整齐的灌木

图 49：
到了冬天，落叶树的枝杈特点就会显现出来

特征与形态
的平衡

  了解植物在不同观赏距离下的视觉效果是非常重要的，因为在景观建筑学中，观赏距离是从能看到景观的位置算起的。一个形体的视觉效果会随着距离的变化而变化。在远处，人们的眼睛能分辨的只是一个剪影而不是形体。中距离时，植物会在阴影效果的作用下变得更立体。近距离观看时，植物的色彩和质感在视觉上会比植物的形态更为突出。在确定具体需要采用多少种植物时，应该把距离的因素考虑进去。因为，即便是非常细心的人在远处也只能从一大片树林中辨别出有限的几个品种。

特征

  与外形一样，植物的特征也大多体现在外观上。它们反映了植物的外观在生长形态上的特点。在景观建筑学中，树是众多特征最为突出的植物之一。对于树和灌木而言，它们的特征在冬天可以看得非常清楚。〉见图 49 当我们以形态特征为依据对植物进行分类时，如采用图示的方式来表现，它们看起来就如同被修剪过一样。〉见图 50 对于什么样的植物在视觉上适合种在什么样的环境中的问题，以外观和特征为依据对植物进行分类可以帮助我们建立一个大体的印象。树冠规整密实的树适合用在秩序井然的场合中，在这里，树都是等距种植的（如城镇广场）。而树冠疏松、形态自由的树可以让刻板生硬的建筑立面活跃起来。

〉✎

	形态	举例	用途
	圆形, 球形	法国梧桐 (成年植株)	适用于讲求秩序的场合,采用单排、双排或是矩阵式种植形态
	圆形, 蛋形	尖叶槭 "Cleveland" (挪威槭)	适用于城市中的开阔场所,包括广场、街道和公园等
	不规则形 树冠松散形	三刺皂荚 (美国皂荚)	适用于相对自由的场合,适合在混合种植区中单独种植
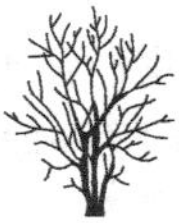	多树干形	鸡爪槭 (日本红枫)	适合种在建筑旁边,起到点睛的作用
	锥形	榛树 (土耳其榛)	适合群植或是种在其他植物之中起到视觉焦点的作用
	柱形	钻天杨 "Italica" [黑杨(伦巴第杨)]	适合用在开阔的景区地势平坦或略有起伏的开阔场地中,用来突出线性元素(林荫道)。与明显的水平造型元素和入口区形成对比
	下垂形	垂枝桦 (银桦)	具有形式感的单株大树,适合单植或是松散成组布置,适用于风景名胜公园或是造型复杂多变的建筑

图 50:
树种的不同特性

　　草及草本植物也同样有着鲜明的特征。花、叶、茎以及新芽的生长方向等都有着不同的生长形态：对于只有一个新芽滋生点的草本植物而言，其叶片生长在贴近地面的位置并且只会长出一枝开花的茎干，例如毛蕊花（毛蕊花属）或是洋地黄（洋地黄属），而挺直生长的丛生植物则会径直向上生长，例如鸢尾和中国芒草（芒属）。斜向生长的丛生植物则会向四面八方伸展出许多柔和的弧线，例如萱草（萱草属）和狼尾草（狼尾草属）。不同的生长形态其效果也各不相同：挺直向上的草本植物给人一种突出有力的感觉，而斜向生长的草本植物则会显得柔和而优雅。当我们需要把多种不同的植物搭配在一起时，它们能否在设计的角度上很好地组合在一起，我们可以从它们的生长特性和生长形态中作出判断。〉见"植物也是一种材料，设计原则"一章

　　植物特性的形成主要是光照和生存竞争的结果。生长在背阴处的喜光植物就无法形成正常的特征；最终会造成植物发育不良而且不会开花。

质感　　质感是植物最有形式感的特性之一。无论是整株植物的生长密度，还是单片叶子、茎干以及嫩芽的表面特质都能形成质感效果。所谓质感就是指植物叶子的整体特征：诸如单片叶子的形态及其表面特质、叶子的规格、排列方式、数量以及叶片表面对光线的反射效果等。纤弱的嫩枝和嫩芽也能产生一种质感效果。简单来说，质感效果可分为"非常细"（草坪），"细"，"适中"，"粗"以及"非常粗"等几类。〉见图 51 修剪后的绿篱与精心修剪的草坪效果类似，通常而言，它们的外观都是密实、精细并且"平整"的，表面看起来很协调，像是一面墙。例如，经过修剪的紫杉绿篱所具有的那种致密精细的质感以及犹如建造出来的外观造型会给人们留下规整严谨的感觉，而自由生长的玫瑰花丛则突出了自然天成的感觉。如果要让植物与建筑物或其他构筑物产生呼应，那么在植物质感的设计中就需要把建筑现有的材料质感、拟采用的材料质感以及建筑的造型考虑进去。如果

图51：
不同质感的示例：细、中细、中粗、粗

图52：
草格形成的肌理结构

图53：
鲜花组成的图案仿佛一张铺开的地毯

植物的叶片大小都一样，犹如墙上的砖块一样，那么看起来就很容易让人们觉得单调。〉见"植物也是一种材料，设计原则"一章植物的质感有着多种作用：

——可以让植物群的外观显得连贯而有力；

——可以起到强调的作用；

——精细的质感效果可以营造出清晰协调的背景，可以在视觉上扩大花园空间；

——可以用来配合突出景观的纵深感；

——可以让种植区形成统一的外观形象，只需让不同的植物形成一条相同质感的连续植物带。

图54：
富有韵律感的草带形成了一个连续结构

图55：
种植打破了地面的组织结构

组织结构　　　组织结构是指一个设计单元内在的构成方式。组织结构是由于内部元素的重复出现而形成的。对"组织结构"的体现适用于所有层面的设计。为了便于人们理解，无论是种植展示区、设计草图还是说明文字都需要具备类似组织结构。

表面组织
结构　　　利用大量相同或是相似的图形可以在表层上形成某种构成效果。〉见图52 规则的组织形态具有装饰性和图案化效果（壁纸、表面涂漆的织物、地毯等），重点突出的是平面感。〉见图53 而不规则的组织形态则会显得更加生动也更有立体感。〉见图54、图55 材料及植物的质感可以在平面上营造出诸多不同的效果。例如，由同一品种的草本植物所构成的花圃可以塑造出强烈的视觉效果。〉见图56 及图57 而由多种多样的夏季花卉和草本植物构成的地被型花圃则会形成另外一种平面形态效果。绿化展示区的内部结构是可以人为创造的，例如，通过不断重复使用叶片形态鲜明的植物（如草或者蕨类植物）即可实现。〉见"植物也是一种材料，设计原则"一章

空间组织
结构　　　空间结构既可以是通透的，也可以是封闭的。在长满大树的森林中，人们会发现自己置身于一个空间结构之中。人们的前后左右都是树，头顶则布满了树杈和嫩枝。正是由于这些为数众多的相同元素或是相似元素的存在，再结合它们划定的区域就共同形成了一个空间结构。〉见图58 及图59 要想让一个花园成为一个空间结构，"一个架构"，就要选择相同的或是相似的空间限定元素（如树或者木本植物造型），并且反复地将它们整合到这个空间中去。种植效果可以是浓密

图 56：
由统一的鲜花铺成的表面结构

图 57：
由坐垫般整齐的灌木带形成的表面结构

图 58：
由成排的竖向植物体量形成的空间结构

图 59：
由重复的树列形成的空间结构

的、通透的、均匀的、有韵律的或是自由的，形式多种多样。〉见图 60、图 61 及"植物作为建筑材料，设计原则"一章对于掉光叶子的落叶木本植物和某些针叶木本植物而言，它们的空间结构和枝杈特征是显而易见的。这些枝杈所形成的线条和图案可以很好地起到底衬的作用。〉"植物也是一种材料，设计原则"一章

轮廓线　　　　植物的外形线或是剪影线被称为植物的轮廓线。木本植物之间的一个显著差异就在于它们的轮廓线有的是完整的，有的则是不完整的。而修剪整齐的木本植物及绿篱由于会定期进行修剪，因此它们都有着致密的质感和清晰连贯的直线形轮廓。它们是塑造花园形态的重

图60：
植物之间自然流畅的融合

图61：
不同类型的植物构成了一个纵横交织的空间

要元素。〉见图62 在几何形态的花园中，如果缺少修剪成形的木本植物那简直就是不能想象的。质感致密并且形态自由的木本植物与修剪成自然形态的木本植物一样有着连续清晰的形象。它们的塑性造型和视觉形象给人们一种"沉稳感"。〉见图63 在不完整的轮廓线中，有些植物的叶片也是有秩序的，例如那些有着层叠枝条的植物（如灯台树，*Cornus controversa*）或是层次统一的植物（塞尔维亚云杉，*Picea ormorika*），其他则是不规则的、松散的。观察距离越近，人们就越是能够清晰地分辨出单片叶子的外形。

色彩　　　　　尽管植物外观的特征和形态是最为重要的视觉元素，但是绝大多数人在花园中只会留意花、叶和果实的颜色和质感。虽说不同植物的色彩和质感凸显了季节的变化，但是一个花园逐渐成熟的过程却是通过植物的形态和特征来体现的。植物的色彩千变万化，而叶片及花朵的光影效果及丰富的表面质感（光亮或是粗糙等）也会增加色彩的变化。色彩是可以依据其效果进行系统梳理和归类的，也是可以利用色轮图及色彩表来进行测定的。色调以及色彩的明度和亮度决定了色彩的效果。当我们在景观设计中运用色彩时，一定不要忘记，在绝大多数花园和园林中，绿色是主导色，而秋冬季节的主导色则是棕色，除此之外的其他颜色只占有非常有限的一小片区域。

色彩明度　　　　色彩的形成有赖于光的存在。光的种类、光照强度和入射角度对于色彩效果的影响至关重要。植物在阳光下和在阴影中的色彩效果是截然不同的。在设计中，室外场地的哪些区域应该在一天中的哪个时

图62：
修剪后的绿篱轮廓线清晰而连续

图63：
经过修剪的欧洲山松致密的轮廓线柔和而连续

段能够照到阳光是需要给予明确的。漫射光会减弱色彩的明度，而直射光则会增加色彩的明度。因此，无论是晴天还是阴天都会影响叶子与花的外观效果，人工照明的效果也是一样。在白天，黄色和黄绿色的色彩明度最高。而在夜晚，蓝绿色的明度最高。在强光照射下或是光线逐渐变暗的情况下，尽管颜色还能看得见，但是色彩会逐渐变灰。绿化展示区的景深感也会随着光照方向的改变而发生变化。与午间的光线相比，早晨和傍晚的阳光（光线来自侧面）可以在花园或是景观中营造出更加强烈的空间景深。而漫射光线营造出的景深效果则会相对较弱。

每一种色彩都有着自己特有的明度。蓝色显得暗，黄色显得亮。红色亮度适中但与橙色相比则亮度略暗。色调会因黑色与白色的加入而发生变化。色调的梯度变化可以通过色轮来表现。〉见图64 纯度最高的光谱色位于色轮的最外圈，而最内圈则代表了物体表面亮度最高时的"中性"白和物体表面亮度最低时的"中性"黑。在外圈色和中性色之间的区域，色彩饱和度呈梯度变化。如果蓝色往黑走就会变"重"，它的那种轻灵气质就会消失。如果黄色往白走，它就会逐渐丧失光彩而变得苍白。花的颜色越灰，它的鲜艳度就会越弱，看起来就会觉得越远。因此在视觉上，纯色要比杂色会显得距离更近。

白色能够突出所有颜色的色彩效果。白花植物及杂色（白边、白斑）叶片可以让绿化展示区的背阴区亮起来。银灰色植物也具有类似的亮化作用，尤其是在和白色植物种在一起时。

143

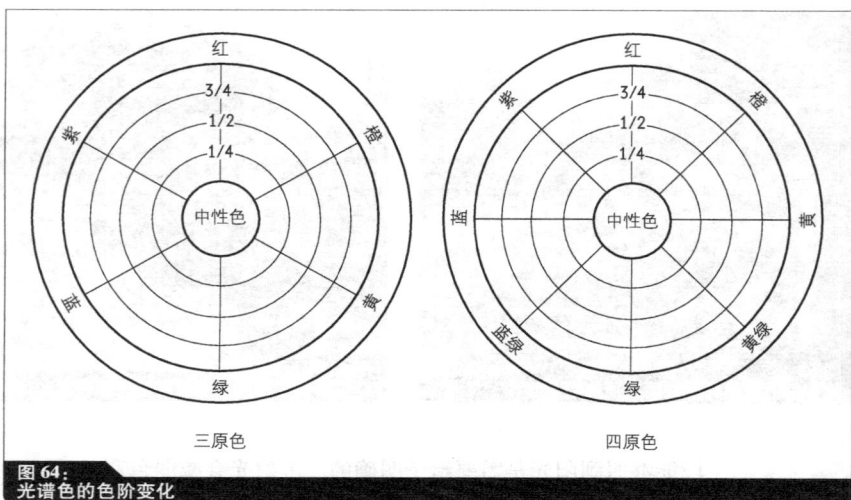

图 64：
光谱色的色阶变化

中性色 · 三原色 · 四原色
红 · 橙 · 黄 · 绿 · 蓝 · 紫 · 黄绿 · 蓝绿
3/4 · 1/2 · 1/4

补色

      互补色位于色轮上的相对位置。绿色、紫色和橙色等间色是由红黄蓝三原色混合后得到的。这六种颜色是白光分解后得到的光谱色。原色总是与间色相对。例如：红色对应着绿色。光在叠加时，彼此相对的两种颜色混合后就变成白色。补色可以彼此突出对方的色彩效果和色彩饱和度。红色在绿色背景下会显得尤其鲜艳，黄色在紫色背景下以及蓝色在橙色背景下都是如此，反之亦然。每种色彩都具有让其他颜色向着自己的补色方向偏色的作用；绿色会让黄色看起来发红——也就是说，绿色在黄色中生成了自己的红色补色。通过补色的对比作用，黄色中增加了并不存在的红色并会因此偏橙色；蓝色会让绿色看起来发黄；绿色会让蓝色发紫；而黄色则会让绿色看起来发蓝。

      对于红、黄、蓝、绿四原色而言，这几种能够产生最强烈色彩效果的补色在色轮上正好位于两两相对的位置。〉见图64

冷色与暖色

      色轮上的颜色可以分为冷色与暖色。红、橙、黄让人们觉得温暖，黄绿和浅叶绿也是如此。蓝色、蓝紫色和蓝绿色则是冷色。暖色会显得离人近，而冷色会在视觉上有后退感，冷色会使得室外空间的景深看起来要比实际更远一些。中绿和蓝紫被认为是中性色。因此景观的绿色具有镇静和稳定的效果。全部由暖色构成的色彩搭配和全部由冷色构成的色彩搭配都会显得协调统一，但是冷色与暖色的组合会形成对比，从而造成完全不必要的不和谐。

色彩的协调

　　当我们要在一个和谐的环境中使用色彩时，我们可以从这个环境中选出一种颜色，然后采用与该颜色最接近的渐变色即可，例如可以往暖调子走也可以往冷调子走。背景、环境、生长形态、周边植物的质感以及自身的色彩变化都会影响色彩的协调度。木本植物的绿色千变万化，这就需要我们有意识地去设计它们的组合，我们的目的就是要让植物能够形成良好的色彩搭配效果，从而让绿化看起来非常协调。将暗色的针叶木本植物同亮色的落叶木本植物组合在一起可以让色调更加突出。在传统的日式花园中，协调统一的色调是通过大量的常绿树木与灌木来形成的。其他的色彩也只是春天的樱花和秋天的秋叶，这些颜色的效果之所以让人们觉得恰到好处，就是因为它们的存在时间非常短暂。

　　能够彼此协调的两种颜色都是在色轮上处于相对位置的颜色（补色）：橙与蓝、金黄与群青、橘红与蓝绿等。〉见图65 能够相互协调的三种颜色则是在色轮上彼此相隔三分之一距离的颜色：蓝、黄、红或者群青、橘红和黄绿。此外还有另外一种可能的组合，就是在相互协调的两种颜色中加入在色轮上与这两种颜色其中之一相毗邻的颜色（如群青、橘黄、黄），或者从两组相邻的互补色中去掉其中的一种（如蓝绿、橘红、橙）。两种原色与一间色的组合效果会非常强烈（红、蓝、紫）。而两种间色与一种原色的组合则会产生非常微妙的效果（绿、橙、红）。尽管在理论上存在着非常多的协调色——但并非任何一种红色与任何一种绿都能相互协调。如果两种颜色的色彩不协调，那么可以借助第三种颜色来弱化这种冲突。但是在相互协调的三种颜色中，再增加任何一种其他的颜色都会很难再塑造出富有表现力的形象。

示例：

　　当花圃中花卉的颜色只有一种时称为单色构图。在只有一种颜色的花圃中，形态的对比就会变得非常突出，例如将大量的蓍草（Achillea）、一枝黄花（Solidago）以及黑心金光菊（Rudbeckia）密植在一起。

旁门：

　　即便只有一种颜色也可以通过光线、明暗以及色彩的冷暖变化等方式形成丰富的色调。将基础色调控制在有限的几种之内即可形成内部的协调统一，并且可以避免出现唐突或是凌乱的色彩。强烈的颜色在日照充足的区域中效果最佳，而浅色尽管比较低调，但是在阴影区中效果也会非常突出。

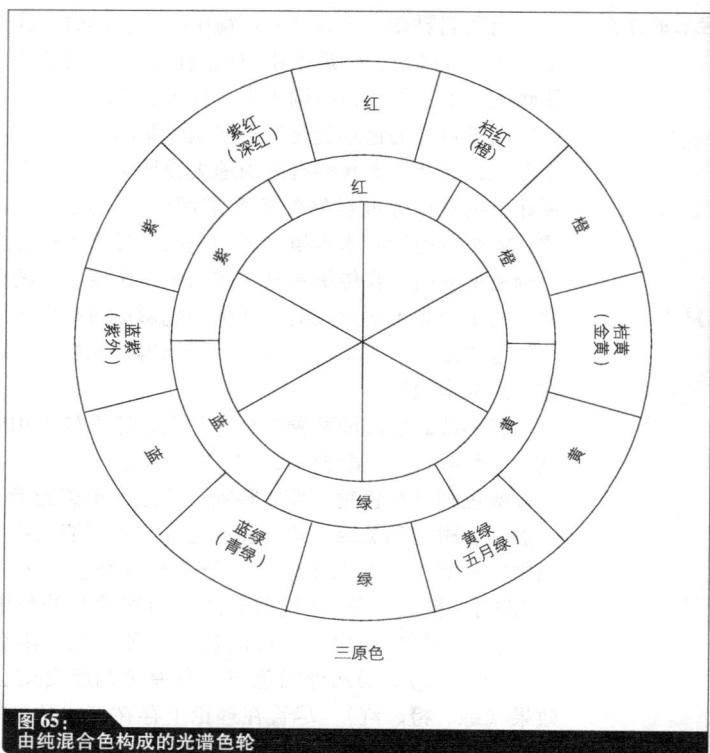

图65：
由纯混合色构成的光谱色轮

（色轮标注）红、桔红（橙）、橙、桔黄（金黄）、黄、黄绿（五月绿）、绿、蓝绿（青绿）、蓝、蓝紫（紫外）、紫、紫红（深红）；内圈：红、橙、黄、绿、蓝、紫；三原色

　　在色彩组合中加入银灰色的叶子可以获得非常突出的色彩协调效果。在这种情况下，诸如红色和蓝色这样的纯色就会显得尤其鲜艳，而低纯度的颜色和清淡柔和的颜色也能展现出它们的最佳效果。〉见表5

表5：
协调色示例

双色协调	三色协调
蓝—橙	蓝—红—黄
橘黄—群青	蓝—红—银灰
橙—银灰	浅蓝—黄—银灰
玫瑰红—银灰	黄—白—银灰

时间力学

以植物为素材的设计超越了二维和三维空间并将四维空间的时间也囊括进来。植物与混凝土和石材的不同之处就在于它们是一种有生命的材料，伴随着植物的生长，它们的外形也会不断地发生变化。这种变化的速度有时会非常明显。甚至每天都能看到变化的发生，特别是在长叶、开花和结果的时期。生长在人类生活的纬度范围内的植物会随着季节的更替常年发生着周而复始地变化，这种变化会依据植物寿命的长短可以持续几十年，甚至几百年。在花园中，人们可以观察到一个连续的生长和逐渐死亡的过程。但是，这种与生俱来的变化也会带来一些问题。例如，一座花园到什么时候才算是完全建成？到什么时候才算是初现成效，又在什么时候开始丧失品质等。以植物为素材的设计还意味着我们需要等待很长的一段时间才能看到效果，因为植物的生长需要时间。与建成多年的花园相比，在空地上刚刚完成的种植布局会看起来空空荡荡像是没有完工的样子。如果设计师或是使用者没有考虑到时间的因素，他们必然会觉得十分失望。因此在实施这样的设计时，最好能选用一些与空间比例相匹配的规格大一些的植物，以便在初期阶段就能形成一定的空间和格局。此外还有一个适用原则：植物的形态和特征也能塑造室外场地的格局，而不同植物的色彩和质感还可以突出季节的变化。

季节变化　　　尽管很多木本植物的空间结构都不会发生太大的变化，但是在初春和入秋时节，植物外观颜色的变化通常都是富有戏剧性的。〉见图66 任何一种植物都有自己特有的一系列季相变化。而对于很多常绿植物和落叶木本植物而言，它们的变化则会历经多个季节才能显现出来。在夏季，落叶乔木构成了绿化展示区的主体框架，但是到了冬季，常绿植物和针叶植物就会在视觉上跳出来，如果种植得当，它们同样也能构成绿地的框架。这类植物的外观变化非常小，因此能够形成一种稳定感。以杜鹃为例，它们在五月繁花怒放，但是到了夏季却很少有人注意到它们，入冬之后，由于它们的叶子是常绿的，因此又会再度进入人们的视线。夏绿型落叶乔木在叶子落光之后就会变成线条和图案。〉见"植物也是一种材料，植物：外观"一章草本植物展示区的外观变化尤其明显。进入冬天，很多草本植物地面以上的部分都会枯萎死亡，但是到了春天又会在原地长出新芽，并且迅速长高长大。因此在选择植物时，建议要考虑到它们在一年之中的外观变化，特别是对于那些一年四季都能看得见的花园，比如说私家花园，尤其需要注意这点。

图66：
草地上散布的郁金香营造出一个繁花似锦的醒目场景

在构思植物设计和种植布局时，很重要的一点就是要让色彩从早春到晚秋延续不断。一个方法就是以花期为依据对植物进行分组，并布置在场地或是景观中的不同位置，如果将多种花卉同时栽种在同一个位置则会削弱整体印象，而且还会显得设计很没有想法。在自然景观中，姹紫嫣红的时期通常都很短暂，在这之后就会进入平淡期。如果想要得到"色彩喷涌不绝"的效果，我们可以利用植物的应季色彩来实现。为此，可在春天选择开黄花和蓝花的植物，夏初采用深蓝色、白色或是粉色花卉，夏末选用大红、深红、紫色以及深黄色花卉。棕色和紫色的叶子或花朵适合用在秋天，深绿和棕色的叶子以及红色的浆果适合用在冬天。

窍门：

建议抓住每一个机会去研究植物在一年之中不同时间里的外观形态。在苗圃、草本植物花园、植物园以及教育和绿化展示区中，很多植物都配有名称标牌。仔细观察那些"常见的"和"不常见的"植物品种，分析它们的外观和感官特质，借此逐渐建立个人的判断力，这对于形成个人理念和手法是非常必要的。

注意：

设计过程中一定要特别注意的是，从应季效果的角度选择的植物也要能在一年中的其他时段对室外场地的布局设计有所贡献。

148

图67：
修剪过的绿篱在全年之中都能塑造花园的空间

图68：
春天迷人的樱花林荫道

应季效果和
景观形态

在设计应季效果时非常重要的一点是，很多植物虽然花开艳丽，但是对于塑造花园的形态却没有什么帮助。例如，丁香的花很漂亮，但是它的叶子和枝干的形态却乏善可陈。形态不佳的叶子和枝干所带来的影响可以通过强化总体视觉效果来予以弱化。仍以丁香为例，可在其前方栽植低矮浓密的绿篱或是经过修剪的绿篱，或是让丁香给其他植物做背景，这样一来就可以在丁香不开花的期间借助这些植物的季相特点来增加变化。杂交茶香玫瑰也同样是花很漂亮但是枝叶形态不佳的例子，剪枝之后情况更甚。因此玫瑰园通常都布置得比较规矩，以便更好地突出玫瑰的特质，此外通常还会在其外围栽植一些低矮常绿的修剪绿篱，从而在视觉上弱化玫瑰枝干和叶片光秃秃的形象。〉见图67 当玫瑰开花的时候，五颜六色的花朵会在围合绿篱的衬托下显得更加鲜艳夺目。同样，在玫瑰花圃里间隔种植的草本植物也能突出花圃的效果。当然，这些植物只是玫瑰的陪衬，不能让它们喧宾夺主。

很多形态矮小的树却有着悦目的花朵，例如观赏樱花，只是它的花期比较短。因此在设计小规模的花园或是场地局促封闭的花园时，如内庭院，就可以考虑采用这类植物。当它们与高低错落的其他种类的树和灌木种在一起时，它们就会不时地在环境中的某个地方迸发出一片短暂的色彩。为了进一步更好地表现出季节的色彩变化，我们还可以把大量能开花的树栽种成整齐的单排、双排或是矩阵的形式。〉见图68

图 69:
地方园艺展览会上用花圃构成的花毯

花圃

设置花圃的主要目的是为了获得强烈的色彩效果。一年生的不耐寒的夏季花卉每年都需要重新栽种或是培育。从前期准备、着手栽种到后期维护（需要不断地浇水和施肥）都需要消耗大量的人力和财力，但是对于那些精心选定的场地、雄伟壮丽的场所或是人们经常前往的地方，诸如雄伟建筑的前广场、公共空间、步行区、历史久远的花园或公园、市民公园以及一些特殊的花园等，付出如此大的代价也是值得的。〉见图69 无论是在简朴的乡村、农场花园，还是种在花盆或是盒子中做装饰，一年生的植物通常都会是主角。设计出色、维护精心的花圃对于打造一个城镇的美好形象发挥着巨大的作用。

用春、夏、秋三季的花卉组成的花圃可以营造出季相变化。在风景区中也可以设置这样的花圃，只要设计能够与局部景观或者区域景观取得呼应即可。不过，需要依据天然植被的组成（例如森林或是石楠花等）来选择和确定花卉植物的颜色和布局。夏末对于公园和景区而言是一个相当平淡的时期，但是可以通过栽植在这段时间开花的一年生植物来增加景观的趣味性。

秋季

在冬季到来之前，由树和灌木的叶子颜色所形成的秋天的色彩是最受人们喜爱的一种色彩变化。为了获得最为强烈的视觉效果，可以将有季相效果的植物与常绿植物、针叶植物或是到了年末叶子才会变色的植物搭配在一起。这样一来，在偏冷的绿调子的托衬下，暖调子的秋色会更加醒目。那些仍然留在树上或灌木上的叶子以及散落在小

图 70:
霜凸显出了植物的轮廓

图 71:
修剪过的法国梧桐在冬季的奇特形象

径和草坪上的叶子都有着温暖鲜艳的黄色、橙色、红色和深红色，它们让花园和公园呈现出一种印象派效果。如果不加人为干涉，那么在每一棵独立大树或是单株木本植物的下方地面上都会在很短时间内形成一大片美妙的色彩。那些能够结出鲜艳浆果的植物在深秋和初冬并不太突出，除非有常绿植物给它们作背景。要想达到一定的视觉效果，就需要使用大量变色时间晚的植物和常绿类植物来衬托出浆果的颜色。而在小尺度空间中，例如花园中，也许只需一棵树就能达到同样的效果。

冬季

在冬天，植物的质感主要靠霜和雪来突出，特别是那些纤巧的植物形态诸如草类、蕨类以及结有果实的草本植物的枝茎等。〉见图 70 也正因如此，植物的这些（遭受霜冻的）部分不得不在开春之前被剪掉。总而言之，与其他三季相比，冬季只有一小部分植物能够形成视觉吸引力。对于由常绿植物和各色浆果形成的冬季效果而言，由于它们并不能持续整个冬天，因此某些植物品种的彩色枝条就会显得弥足珍贵。例如，山茱萸科的"西伯利亚"红瑞木就有着醒目的红色枝条，它们在冬季会让景观显得别具一格。当它们与白色树干的桦树和常绿木本植物种在一起时，效果还会进一步得到加强。当建筑的窗框也是同样颜色时，这些红色枝条的效果就会尤为突出。有着黄色枝条的金枝梾木（*Cornus stolonifera* "Flaviramea"）也具有同样的效果。某些种类的树和灌木的枝权结构也能在冬天形成一定的视觉吸引力。〉见图 71 在进行植物选择时可以这样来考虑：为了得到好的图形效果，植

151

图 72：
在路边种上小树可以通过对照向人们展示出树的寿命和生长趋势

物在很多角度都是以天空为背景的，要么也会选择一个相对简单的背景，例如一堵墙或是建筑的外墙面等。〉见"植物也是一种材料，植物：外观"一章矩阵式种植的成年大树之间彼此交织的枝杈在天空的衬托下会形成另外一种视觉效果。设计中一定要考虑到冬天的影响，这一点经常会被人们所忽略。

体形的发展
与生命周期　　植物的外观在其一生之中是会不断发生变化的。它们的生长速度以及随之带来的形态变化都是受它们所属的下述生物群落所决定的：

　　—树；

　　—灌木；

　　—草本植物；

　　—球茎类植物；

　　— 一年生植物及两年生植物（夏季花卉）。

树　　　树的寿命很长，但生长速度相对较慢。〉见图72 它们的生长会历经季节的更替延续上几十年。在花园或是景观之中，树是最重要的也是最"稳定"的元素。它们提供了时空的延续性和存在感。树可以将城市与周边环境联系起来，还可以将建筑与城市的各个区域联系起来。尤其是在城市开放空间的植物设计中，一定要考虑到植物生长速度较慢的问题。当树冠下方的草本植物、灌木以及草坪受到过多的遮蔽时，植物的根系就会展开水分与养分的争夺，而阴影区内的植物就会出现生长不良或者死亡的情况。

灌木　　　　　　灌木也会年复一年地慢慢长大，但它们却没有树的寿命长，也长不到树那么大。它们可以用来塑造地表形态或是用作分界线。〉见"空间结构，边界"一章灌木可以在树与地被植物之间起到联系过渡的作用；一个只有树、草本植物和草坪的公园会显得非常空旷，几乎没有空间景深。灌木可以作为公园与花园之间的过渡或是公园与景区之间的过渡。

草本植物　　　　草本植物是多年生植物，它们与树和灌木的不同之处就在于，它们的地表以上的部分会在秋天过后枯萎死亡。到了春天，草本植物又会从它们耐寒的地下贮藏器官中长出新芽。草本植物的生长高度从地毯毛绒那么高一直到 2 米以上。草本植物随着季节更替所发生的形态变化给花园或是室外空间带来了巨大的活力。

球茎植物　　　　隐芽植物（Geophytes）——长有球茎、块茎、根茎一类的地下贮藏器官的植物——它们在一年之中的大部分时间里都埋在土壤之下不为人知。年初之际，当树和灌木的叶子尚未发芽，阳光仍然可以照射到地面上的时候，它们中的大部分就会纷纷现身。它们的自然栖息地位于森林之中或是森林的边缘区域。它们的叶子寿命很短，而且在开花之后叶子就会停止萌发。叶子的停止生长对于这类植物的健康和开花能力而言十分关键。为了让它们更好地保护叶片，应该把它们种在低平的植物当中而不是草坪的表面。通常它们的寿命也比较长，而且不存在跟其他植物在外观或是生存上出现竞争的问题。但是花期比较晚的球茎植物（如郁金香）大多来自干旱地区，由于那里只有稀疏的植被，几乎不存在生存竞争的问题。正因如此，它们不适合跟草本植物和灌木种在一起。适合种在草坪上的隐芽植物有番红花和水仙等。当它们的叶片刚开始变黄时就应该及时给予摘除。

一年生及两
年生植物　　　　一年生植物只能存活一个生长季。它们通常用作临时展示。如果它们没有表现出竞争性行为，就可以与草本植物种在一起。

　　　　两年生植物可以存活两个生长季。它们通常会在第二年开花并在死亡之前结出很多种子。

P71　　　　　　**设计原则**

　　　　为了取得好的种植设计效果，植物的外观差异——大小、形态、颜色、质感——必须统一起来形成一种内在的联系。这就需要有一个统一的理念，一个主题。主题理念构成了设计的内容，可以通过空间、植物和材料来给予落实。设计通则方面的知识，诸如对比、均衡、重复、节奏以及秩序等，可以帮助我们把理念表达得更清晰、更易于辨识，还能让理念从众多园林书籍的经典种植案例中脱颖出来。

对比　　　　对比手法是最为重要的植物设计原则之一。通过制造矛盾冲突和吸引力来唤起人们的兴趣。对比会让差异变得更加明显。当至少两种相反的效果碰到一起时就会形成对比。在开满鲜花的草坪上穿行而过的一条修剪出的小路就是一个简单的例子。在自然景观中，可以看到很多诸如这类的例子。山毛榉树林到了春天就会在地面上开满十分醒目的五叶银莲花（Anemone nemorosa），它们与山毛榉光秃秃的粗大树干形成了对比。

利用形态、大小以及色彩的对比在植物之间建立起的这种人为关系是用来突出个体植物的重要手法。强对比效果，如色彩对比，无需特别关注就能立刻引起人们的注意。弱对比效果，例如质感对比，就需要多花一些时间和精力来观察才能有所察觉。当种植区中的植物被设计成人们每转过一个拐角就会发现一种新的对比效果时，这样的设计就会给人们带来惊喜。

对比也需要讲究平衡。采用安静的背景，诸如建筑墙面、冷色调或中性色调的植物颜色（绿、灰）以及采用逐渐过渡的手法来处理植物的高度和色彩变化等，都能突出对比的均衡性。相对于大型植物和色彩鲜艳的植物而言，小型植物以及色彩低调的植物就需要以数量取胜。过多的强烈对比会让人们感觉疲倦，而太过相似或者不够清晰的对比则会显得意兴阑珊、枯燥无味。适合用在植物设计中的对比手法如下：

——生长形态对比；

——质感对比；

——色彩对比；

——明暗对比；

——图底对比；

——虚实对比；

——光影对比；

——正负对比（凹/凸）；

——"阴阳"对比。

生长形态　　生长形态的对比可以突出植物的动态及静态效果。通过设置对比
对比　　　元素，植物的生长形态及其特质都会比单独种植时表现得更为突出。〉见图73 对比双方只有在体量上大体相当的时候才会最易于被人们察觉。〉见"空间结构，比例"一章适合采用对比手法的形态包括：

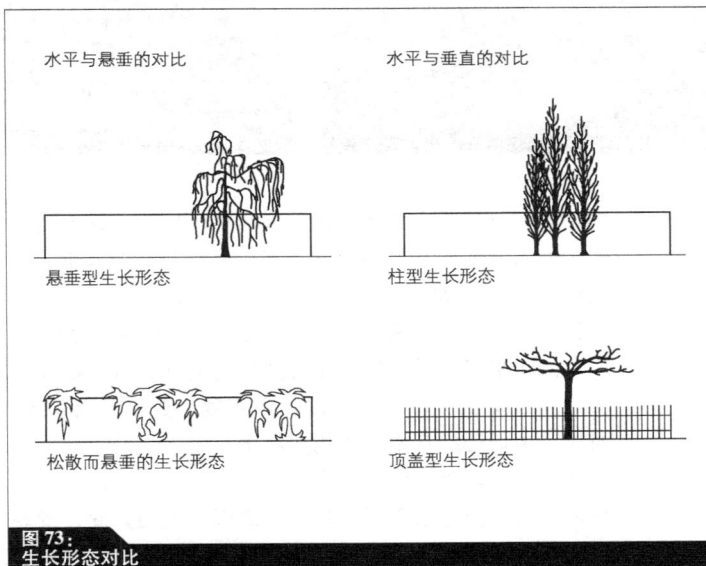

水平与悬垂的对比　　　　　　　　水平与垂直的对比

悬垂型生长形态　　　　　　　　　柱型生长形态

松散而悬垂的生长形态　　　　　　顶盖型生长形态

图73:
生长形态对比

　　——垂直形对圆形、无方向型；

　　——水平形对自由型；

　　——悬垂形对倾斜型；

　　——松散形对紧密型、圆形；

　　——自由形对规整型；

　　——直线形对无方向型，圆形；

　　——直线形对平面型；

　　——装饰造型对艺术造型。

　　圆形是没有方向的，有一种静态感，可以采用多方向生长的悬垂形态来与其取得对比效果。一条弧形植物带或是沿着弯曲的小路栽种

图 74：
生长形态的对比

图 75：
质感对比

的球形植物都具有这个效果。在自然环境中，蜿蜒的河床上散布的不规则卵石及河砾石所构成的形态也具有类似的效果。线性叶片（如草类或鸢尾）与大而圆、扁而平的叶片（如玉簪、睡莲等）可以形成对比。由水平枝干和宽大的伞形树冠（美国木豆树、美国梓树）所构成的水平形态的植物或是经过修剪的绿篱等，都能与起伏的地面造型或是竖向的形态（柱状的木本植物、建筑物等）形成水平方向的对比。由低矮的草本植物构成的地表植被也可以形成一个简洁醒目的形态，从而与垂直的树干形成对比。〉见图74 在距离相同的情况下，垂直形态总是会比水平形态显得距离近。正因如此，景观中的柱状形态即便在远处看也会觉得醒目。在起伏的地面上，垂直形态会因对比而显得稳定；如果利用多方向感的植物（斜向的或是悬垂的）来弱化这个对比效果，就会给景观带来动感。从设计的角度来看，生长浓密并且轮廓线连续的树同线条化、图案化的树能够形成一种非常好的对比效果。〉见 "植物也是一种材料，植物：外观" 一章

质感对比 　　质感对比能够把创造力带入植物设计之中。这在那些安静的种植设计中表现得尤为清晰。在植物高度变化丰富的布局中，人们的注意力会被对比鲜明的叶片与植物组织方式之间的相互作用所吸引。〉见图75 质感效果并不会因色彩而被削弱，白花、白边叶片或是彩色叶片还能够强化质感的对比效果。植物的质感对比效果有：

　　—松散的对浓密的；

　　—细的对粗的；

—有光泽对无光泽；

　　—柔软的对坚硬的；

　　—有质感的对光滑的；

　　—粗糙的对光滑的；

　　—精细的对粗糙的；

　　—透明的对革质的；

　　—窄的对宽的；

　　—线性的对无方向性的。

　　粗糙质感的植物给人以稳定有力的感觉，而精细质感的植物则会显得平静和缄默。观察距离相同时，大叶植物要比小叶植物感觉距离人们近一些。〉见"植物也是一种材料，植物：外观"一章

色彩对比　　色彩对比不仅能够让绿化区显得更活泼，还能够强化色彩效果。最重要的色彩对比效果有：

　　—明暗对比；

　　—冷暖对比；

　　—补色对比（色轮上相对位置的色彩）；

　　—品质对比（鲜艳与灰暗的色彩对比配合有光泽与无光泽的质感对比）；

　　—数量对比（不同大小的色彩面积的对比）。

　　最强烈的色彩对比效果是通过双色协调搭配或是三色协调搭配来实现的，也就是使用那些在色轮上处于相对位置的颜色（补色）。〉见"植物也是一种材料，植物：外观"一章不仅花的颜色要彼此和谐，花和周围叶片的颜色（基色）也要取得协调。叶子在刚发芽时的颜色与夏季和秋季时的叶子颜色都不一样，不同品种植物的叶子颜色也各不相同（黄绿、绿、蓝绿、红棕等）。

注意：
　　"少就是多"的设计原则也同样适用于色彩的运用。少可以让设计理念表达得更清晰，也更突出。只选择一种植物、但配合使用该植物的多个品种就是一种能够让设计简洁突出的办法。例如，鸢尾的形态很完美，花也漂亮，而且花的颜色很多，但其剑形叶片的外观却很简单。

当使用其他光谱色来代替绿色作为基底色时，它们中的绝大多数颜色都会在与银灰色搭配时获得很好的效果。在这种情况下，诸如红色、黄色、蓝色都会得到突出，而粉色和其他浅颜色也能够得到充分的体现。在银灰色的映衬下，泥土的颜色显得沉静而突出。灰色叶子的植物通常用在小规模的绿化中，例如，它们可以用来搭配小型柳树或是薰衣草等。白色与任何其他颜色进行搭配都不会出现问题。白色还可以衬托出所有其他的颜色。同样道理，多色组合可以在空间上与白色相毗邻。白色花卉具有活泼、清新、精巧的特质。不过，当白花植物成片种植时（白色花园）要比和多色花卉种在同一个花圃中效果突出得多。当把白花木本植物种在深色的针叶树前或是种在阴影区内时，就会形成很好的明暗对比效果，这和把白色树干的桦树种在深色沉闷的背景前所得到的效果是一样的。⟩见图76

植物的品种、数量以及选用色彩的分布都必须协调均衡。色彩纯度低的植物要想压倒色彩纯度高的植物，就必须在数量上超过后者。在歌德（Goethe）的色彩理论中，颜色的明度用"光值"（light value）来表示：

黄色 = 9

橙色 = 8

红色 = 6

绿色 = 6

蓝色 = 4

紫色 = 3

这些光值可以用来衡量色彩的构成比例：黄色比紫色（9/3）= 1:3，蓝色比红色（4/6）= 3:2。设计场地的面积越大，观察距离对于颜色区域的面积大小就越有影响。

色彩对比效果可以在同一个花圃中实现，而彼此位置相对并且分别只有一种颜色的两个花圃之间也可以产生同样的对比效果。

光影对比　树枝上的光影变换以及地面上的树影变化都非常吸引人。依据光照强度的不同，光线可以形成强烈的明暗层次变化。叶子、树皮、土壤颜色的不同以及叶子特征和枝杈结构的差异都可以形成独特的阴影图案：斑驳的、亮的、暗的、深的、清晰的、柔和的、彩色的、对比强烈的、模糊的等。投在树下地面上的影子在不断地发生着变化。影子的形态可以告诉我们当前正处于一天中的哪个时段。正午的阳光明亮而强烈，影子会比较短，但是到了傍晚，光线是黄色的，也相对柔和，影子也会慢慢拉长，从而让室外空间的立体感更加突出。当人们

图76：
明暗对比

图77：
公园景区里的光影变化

的目光正视光线或者太阳时，就会感觉睁不开眼睛，但是在阴影中欣赏风景就要惬意得多。在不同的季节里，我们对光和影的诉求也会有所不同。在冬天，我们喜欢能够带来温暖的阳光；而在夏天，我们则偏爱凉爽蔽日的树阴。了解阴影在室外环境里的效果和重要性是非常必要的，在室外环境的设计中作出恰当的选择也同样非常重要。

光影的变化会让树木看起来很立体而且很有艺术感。光会将树的形状投射在地面上形成树影。当一片松散树林的一侧受到阳光照射并在另一侧形成或是投下阴影时，会是一个非常迷人的场景。但是，当树丛或者树林的外轮廓线太过完整连续时，树影就会变得单调乏味。树林外轮廓线上的开口或是在树林前面栽种绿植都可以形成大片的光影区，它们不仅构成了轮廓线、活跃了轮廓线，还给人们提供了景观。〉见图77

节奏 　　要想让花园、公园或是绿地形成一定的秩序和结构，重复使用相同或者相似的植物品种及植物搭配就会显得非常必要。但是仅凭简单的重复并不能形成节奏，只能起到把绿地中的某些区域联系起来的作用。节奏的形成或是富有节奏感的总体布局的形成是通过有规律地重复典型的植物元素来实现的；由此在视觉上将远处和近处的区域联系起来。在花园或是公园中，整体布局的统一感可以通过使用不同特点的节奏性元素将各个独立区域整合起来的方式来实现。如果在室外大面积种植有特色的植物品种，它们就会赋予这个空间鲜明的特色，并会因此形成一个主题；我们只需想一下栗树林荫道和玫瑰园就明白了。

主题植物　　在一大片种类繁多但外观也非常出色的植物群中，我们不仅会看花眼，还会忽视它们的存在。这样的形象不够鲜明，也不够协调，而且缺少趣味，也就不会对人们形成吸引力。人类的眼睛对于若干相同或是相似的元素最为敏感，这是因为这些元素更容易被识别，也会由此形成一个组织结构。重复使用主题植物或植物组合的设计不仅在视觉上显得稳定，而且也更容易被人们所理解。人们的目光也就会时常被吸引过来，并停留在这些点或面上。

　　主题植物构成了景观设计的基础和框架。通过布局设计，它们可以将整个景观统领起来。我们可以根据植物的不同层级来确定它们的种植位置、组合方式和重复方式。木本植物的种植形态主要是由大树来控制的。树可以形成一个永久性的框架，它们可以起到将设计场地的其他区域或是周边环境景观联系起来的作用。大树可以考虑种在建筑的旁边，特别是花园里有建筑物的区域，还可以种在场地的转角处或是用作边界种植等。树的大小应该与建筑和场地空间取得协调。〉见"空间结构，比例"一章小型树、大型灌木、单株木本植物等与大树一样都是塑造空间结构的植物。它们的作用是：强调尺度关系，将建筑与花园的其他空间联系起来，并且形成大树与灌木之间的过渡。它们可以考虑种在花园的入口处、房屋或是附属建筑的转角处或是用作边界种植等。灌木和绿篱从属于塑造空间的植物，它们可以在花园中的不同区域用作空间的收尾和分隔。它们在与主题木本植物的共同配合下，可以让一个花园空间充满个性。再次一级的小型灌木和低矮型木本植物可以给木本植物种植区提供形象上的补充。〉见表6

窍门：

　　可以将效果出色的植物搭配记在或是画在小本子上并且随时放在手边。通过悉心的观察、记录和绘制，植物的形象就会牢牢地印在脑海中，还能由此积累下大量的个人知识，这样一来，当自己着手做设计的时候就可以信手拈来了。

窍门：

　　Richard Hansen 和 Friedrich Stahl 撰写的《多年生植物及其庭院栖息环境》是一本草本植物应用方面的权威参考书（见附录，文献），它在如何针对种植场地和栖息环境的要求来选择植物、每平方米植物种植数量的计算、植物的种植间距以及植物的群集性等方面非常实用。

Ia	大型树	主体木本植物，构成空间永久性框架的植物，需要充足的生长空间，单植或群植均可
Ib	小型树 大型灌木 单株木本植物	构成空间框架的植物，外形尺寸大于周边植物，寿命长，单植或小规模群植。在规模较小的花园中排在级别体系里的首位。它们决定了灌木的选择
II	灌木 绿篱	陪衬型木本植物。最主要的是，其最大外形尺寸要小于主体木本植物。选择得当的灌木可以给草本植物塑造种植格局，还可以给草本植物花卉作陪衬。宜单植或成丛栽植
III	矮绿篱 低矮型木本植物 半灌木	起补充作用的木本植物。最大外形尺寸要小于树和灌木。适合栽植在高大的木本植物下方的地面上用做地被植物。可以为草本植物区塑造种植格局

表7：
草本植物分类

I	单株草本植物	效果突出，少量几棵单株植物即可
II	主题草本植物	构成空间框架的草本植物，外形尺寸大于周边草本植物，寿命长，适于大量性栽种
III	附属草本植物	用于衬托主题草本植物的效果，植株的最大外形尺寸要小于主题草本植物，寿命长
IV	补充植物	植株的最大外形尺寸要小于主题草本植物和附属草本植物。它们的生长不应对主题草本植物和附属草本植物构成影响

大面积自然风格的草本植物区可以采用单株草本植物或是主题草本植物来塑造形态。主题草本植物也被称为核心草本植物和框架草本植物，这是因为它们在每年的生长期之中效果都会十分引人注目，而且寿命很长。可以考虑使用那些形态和色彩适宜并且效果突出的高大型品种。用来衬托主题草本植物的附属草本植物需要大面积有规律地种植，因此它们的外观应该比较低调。起补充作用的草本植物可用来塑造表皮效果或是用来覆盖地面。不同品种的草本植物之间的过渡应当是连贯流畅的。在不同主题的花园中，一种植物或者同种植物可以扮演不同的角色。例如，鸢尾在某个花园中是主题植物，但在另一个花园里就成为附属植物。见表7

高度层次

在草本植物的设计布局中，有一种手法是采用高、中、低三种植物来形成一个三层的结构。草本植物的规律性重复种植不能太过死板，因为这会让绿化区丧失活力和趣味。主题植物之间的间距以及某种主题植物的数量应该依据每个层次的种植长度来调整它的种植规模。

图78：
灌木形成的层次和节奏

图79：
自然式园林中，树和灌木形成的层次

图80：
规整的玫瑰园中，3层种植结构体现出的层次

图81：
林荫道让空间具有了建筑感

当我们在花床上栽植低矮植物时，可以适当考虑把它们种得靠前一些或是让种植位置不要太过靠后。中型或是大型植物可以种在靠前的位置，可以借此营造出向前走或是向后退的感觉。见图78采用将低矮的植物种在前区、中型植物种在中间、高大植物种在后面（或者对于那种从四面八方都能看得到的绿地而言，将高大植物种在中间）的均布手法会让种植看起来呆板而且缺少生气。可以考虑采用两层结构的种植组合。让高大的植物以单植或是小群组的方式种在低矮的地表

图82：
矩阵式种植的红豆杉

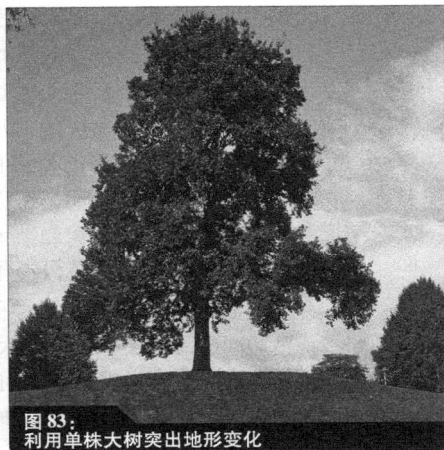

图83：
利用单株大树突出地形变化

植物丛中。木本植物种植区或是木本植物与草本植物的组合种植区也都可以依据已有布局的主次关系设计出层次效果。〉见图79、图80

重复与强调最简单的重复手法就是将完全相同的元素以等距方式排布。由此可以得到一个清晰而且形式高度统一的连续性效果。例如，可以将树布置成单排、双排或是树阵的形式。〉见图81 这些元素可以形成规矩严整的效果。规整种植区的范围最终还可以根据需要任意扩大。

重复可以让选定的植物得到强调并突出它的重要性。重复的元素也可以是植物之间的间隙（例如网格图案），也可以是植物的颜色和质感。〉见图82 利用植物的颜色、规格和质感的渐变处理可以得到一个更具表现力的整体效果，借此可突出种植主题。例如，可以考虑采用同类植物的不同品种或是采用不同陪衬植物的方法等。重要的是：植物品种的选择，特别是主题草本植物的选择，必须严格进行控制，最主要的原因就是，好的植物设计都是简洁清晰的。

更进一步的强调手法还包括利用植物突出场地现状的构造特点和地形特点。例如，规整的树阵可以呼应建筑横平竖直的造型，组树可以用来突出山丘，行道树可以用来烘托街道等。〉见图83

对 称 与 不对 称在轴线两侧的对应位置使用相同的单株植物、修剪植物或是地表植物形态都可以得到对称的效果。道路可以充当对称轴。可使用的元素包括凉亭、格架、藤架以及造型相同或是特征一致的树（例如，林荫道或是修剪成形的木本植物等）。成对栽植的树可以标示出空间的边界、道路沿途发生的功能变化以及与道路相连的大门人口、小桥或是踏步等构筑物。对称手法可以在整体布局中反复使用多次（作为一种装饰）。〉见图84 巴洛克园林中的花坛和视线设计就是采用轴线对称手法的典型实例。在这类园林布局中，地表装饰性的植物形态采用修剪整齐的绿篱围合起来，以此来强化对称效果。对称布局的场地也同样可以采用规整的绿篱加以围合，绿篱的后方则是自然生长的大树和灌木。考虑到人们在途经过程中对景观或是公园的感知方式，可以采用让植物体块持续延伸到景观远方的方式来形成飞逝而过的自然对称效果，只需偶尔让植物体块看起来跟道路对面的另一个植物体块大小相同即可。完全对称的设计会束缚设计师的思路。这类设计手法可以用在形式规整或是威严庄重的花园中，它可以与建筑元素或是装饰性花圃，例如花坛来取得呼应。〉见"植物也是一种材料，时间力学"一章

均衡　　　　　　均衡是设计所追求的共同目标。它是指不同设计元素之间所形成的一种协调平衡的状态。均衡的设计会让人们觉得和谐，而不会像对称格局那么僵硬。可以利用中心构建元素让均衡和对称在同一个景观、公园或是花园中共同存在。轴线两侧植物的精准落位可以形成对称效果，但是在种植时进行细微的调整就可以实现均衡的效果。〉见图85 形象越是突出的建筑，就越是没有必要采用对称种植的方式。可以使用造型、质感或是色彩醒目的植物以特定的间距种植在对称轴的两侧，但是不要让植物之间的组织太过死板。

绘画性设计　　　　绘画性设计是利用不同的视觉元素和元素之间不同的间距所进行的设计处理。〉见图86 既可以采用自由形态或是几何形态，也可以使用这两种形态的组合搭配。植物的布局处理通常会比植物本身更为重要。植物之间的间距以及植物的形态必须要给予精心的甄选，种植的选位也要讲究均衡原则。要让视觉焦点集中在种植区表面的中心区附近。

图84:
对称

图85:
不对称

图86:
绘画性设计

结语

　　园林设计的魅力就在于生命与静止之间的矛盾共生以及植物与空间之间的动态统一。任何有生命的东西都会受到时间和空间的影响。与其他所有艺术表现形式相比，植物景观设计很有可能是最有赖于对时间和空间进行深入观察的艺术形式了。对于一个花园而言，绿化种植只是一个持续发展过程的起始。花园的设计及建造是与园艺学知识分不开的，因为只有经过积极主动的学习才能确保设计师的想法得以实现。要想把植物用好，就需要了解造园知识。它并不仅仅是关于建筑绿化或是野生绿篱一类的简单问题。还涉及相当多的其他内容，诸如英式风格草本植物的种植设计手法或是北欧近些年的一些种植设计手法等，都会给我们带来一些启发。我们在进行素材的选择时，可能会用到以下内容：对简约风格的热衷不要让我们忘记简约其实源于丰富，正是由于丰富，我们的热情才有可能从中作出深思熟虑的选择。如果最初的选择比较少，那么留给人们的印象就不是简约而是简陋了。对于我们的私家花园而言，植物设计的重要性将会是持久的。花园作为我们工作、休闲、娱乐的场所以及它在资源利用方面的表现，它给人们营造出了一个与我们日益机械化和依赖性的社会形成鲜明反差的世界。用心关注植物，而不是漫不经心地去对待它们，这一点尤其重要。这样一来，我们对美的感知会更加敏锐，从而会唤醒我们的内在意识。在我们当今的时代，植物设计是一件非常奢侈的工作，因为它所需要的是我们这个社会最稀有也是最宝贵的东西：时间、精力和空间。对植物的利用反映了我们对自然的感知。当我们将才智、知识和技能重新结合起来的时候，我们也就找到了一个可靠的工作方法来研究环境和它的缩影——花园。

附录

种植平面图

种植平面图是通过图示的方式对设计植物的品种、位置以及数量给出的说明。如果平面图是按比例绘制的，那么就有可能提前计算出场地所需栽植苗木的实际数量，还可以选出适宜的植物搭配组合，这样不仅能够保证种植效果，还能处理好室外设计场地的比例关系。〉见168页图87及103页图7 种植平面图是用来帮助设计师逐渐成熟设计想法的一种工具，最初是绘制在图纸或是电脑上的。在平面布置图中，树是用树干和树冠来表示的，这样一来，树对空间和地面的影响就一目了然了。树冠给出了树的体量，树干则给出了树的位置。根系的范围通常与树冠的大小相当，它们将会占满建筑设施（建筑、地下管线、道路等）周边有限的场地。〉见图88 种植平面图给现场的管理人员及其他相关人员提供了必要的信息来实现设计想法。植物的种类和种植的位置是可以给出估算数据的，绿植及草皮的面积以及实施必要的支护和土壤改良的面积也是可以从设计图上通过计算得到的，园林景观的工程量也是可以估算出来的，而工作计划也就可以开始编制了。

> 窍门：
> 附件中关于苗木和草本植物的清单及表格给出了植物的生长形态以及不同植物每平方米种植数量方面的信息。

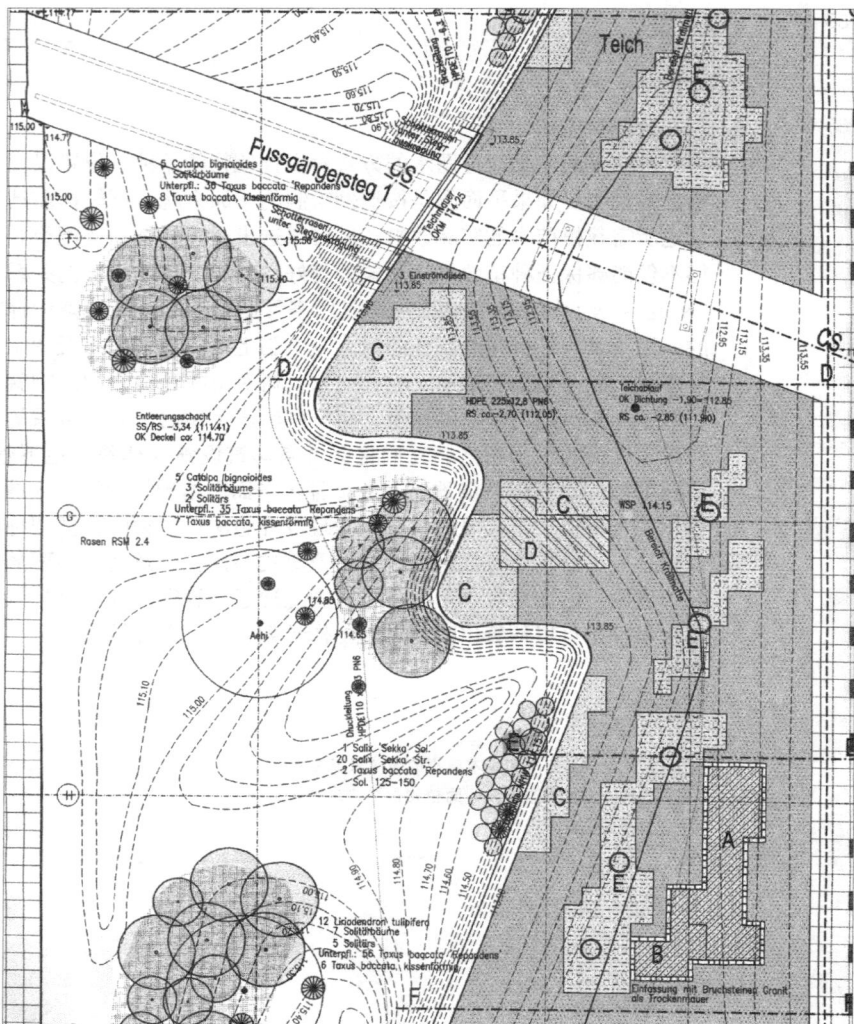

木本植物

- 树的位置
- 规划树
- 修剪成型的欧洲红豆杉 "底衬"
- 种在树下方的欧洲红豆杉 (Repandens)

滨水植物

- C 大叶草本植物 中叶长茎草本植物
- 直立生长的草本植物
- E 芦苇类草本植物
- 竖井区种植平面形态的浮叶型草本植物

图 87：
种植平面图

树冠

树干

根系

树的剖面图

树的顶视图

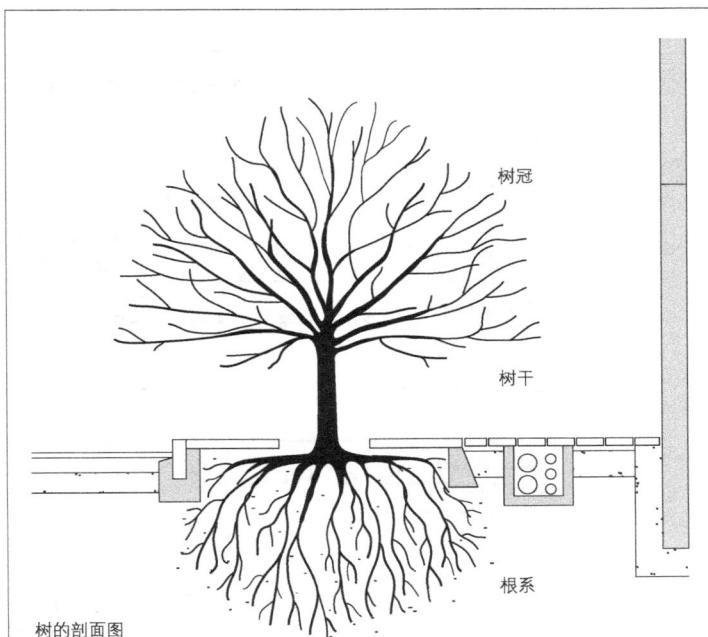

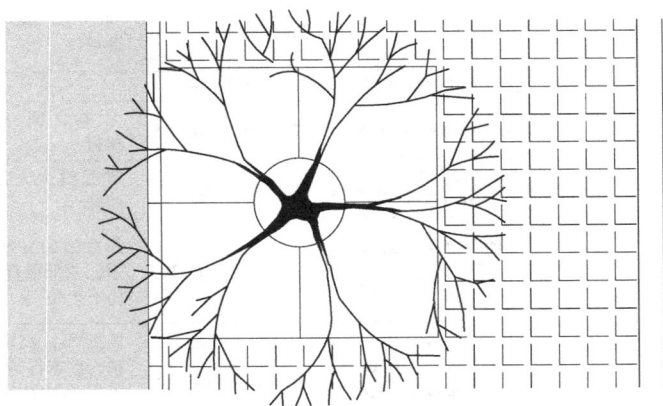

图88：
在设计中，树的整体结构都需要给予考虑

表8：
树的类型及生长特性

学名	名称	高度（米）	宽度（米）	特征/形态	特色
适用于花园及城市空间的小型树					
Acer campestre "Elsrijk"	田园槭	8～12	4～6	致密，圆锥形	秋季变色（黄色），适应城市气候
Acer platanoides "Globosum"	挪威槭	4～6	3～5	致密、球形，随树龄的增加比例逐渐失调	秋季变色（黄色），适应城市气候
Amelanchier Lamarckii	唐棣	5～8	3～5	外观类灌木，横向生长，漏斗形	四月底开白花，总状花序，秋季变色（从黄色到焰红）
Carpinus betulus "Fastigiata"	欧洲鹅耳枥	10～12	5～8	柱状，挺直向上生长	即便未经修剪，树冠也不太大，从始至终都生长得很紧凑
Catalpa bignonioides "Nana"	美国梓树	4～6	3～5	致密的球形	漂亮的大叶片，生长缓慢，不开花
Pyrus calleryana "Chanticleer"	豆梨	7～12	4～5	规则球形	适应城市气候，特别耐热，开白花，秋季变色（鲜红）
Sorbus aria	白花楸	6～12	4～8	大型多分枝灌木或是有着宽大锥形树冠的小型树	九月后会结出漂亮的橘红色果实
Tilia europaea "Pallida"	欧洲椴	几何修剪		盒子形、盖子形或格子形	可进行修剪的木本植物，树冠可以修剪出漂亮的造型
适用于城镇与公园的中冠型至大冠型树					
Acer platanoides	挪威槭	20～30	10～15	大树、圆形树冠	适应城市气候，生长速度快
Acer pseudoplatanus	欧亚槭	20～30	12～15	树冠很大，宽而圆	生长速度快，秋天呈金黄色
Aesculus x carnea "Briotii"	红色欧洲七叶木	8～15	6～10	圆而致密的树冠，主干很直，向上生长	生长缓慢，鲜艳的红色花朵，圆锥花序，几乎不结果
Aesculus hippocastanum	欧洲七叶木	20～25	12～15	椭圆形，高圆顶，致密树冠，可以形成浓密的树阴	开白花，果实很多，秋季色彩亮丽
Ailanthus altissima	臭椿	18～25	8～15	宽大的椭圆形树冠	生长速度快，适应城市气候，无特殊要求
Betula pendula	垂枝桦	12～25	6～8	瘦长的蛋形，枝条松垂	柳絮黄绿色，树皮白棕色，秋季变色

学名	名称	高度(米)	宽度(米)	特征/形态	特色
Catalpa bignonioides	美国梓树	8～12	5～8	伞形圆顶树冠	巨大的心形叶子，15—30厘米长的圆锥花序非常醒目
Corylus colurna	土耳其榛	12～15	6～8	圆锥形树冠，主干连续	适应城市气候，生长健壮，无特殊需求的树种
Fagus sylvatica	欧洲山毛榉	25～35	15～20	巨大的椭圆形树冠	银灰色树干，秋季变色，从黄色到橘黄色
Fraxinus excelsior	欧洲白蜡树	25～35	15～20	蛋形树冠，树龄越长树冠越大，可透过斑驳的阳光	漂亮的羽状叶片，秋季很少变色
Platanus acerifolia	法国梧桐	25～35	15～25	宽大的圆锥形树冠的大树，树龄越长树冠越大	生长旺盛，耐修剪，适应城市气候
Populus nigra "Italica"	欧洲黑杨	25～30	2～5	大型柱状树，树杈和枝条很直，向上生长	生命力强，耐涝
Prunus avium	欧洲甜樱桃木	15～20	8～12	中等尺寸的蛋形树冠	非常漂亮的树，开白花，秋季变色（从黄色到橘黄色）
Quercus robur	欧洲栎	30～40	15～25	起初树冠呈圆锥形，随树龄增加，树冠变大变圆，变疏松	适应城市气候，抗风
Salix alba "Tristis"	白柳	15～20	12～15	中等树，装饰性效果突出，悬垂枝条	具有艺术性效果，树龄越大抗风性越差
Tilia cordata	小叶椴	20～30	10～15	树冠宽大的大树，树冠起初为锥形，后期变为高圆顶形	对城市气候适应度适中，耐修剪
Pinus sylvestris	樟子松	15～30	8～10	具有艺术性效果的大树，造型多变，随着树龄增加，可形成高大的树冠，伞形	双针叶，从绿色到蓝绿色，适应城市气候
Thuja occidentalis "Columna"	香柏	15～20	2～3	中等高度的柱形树	常绿，适应城市气候，耐修剪

参考文献

Ethne Clarke: *Gardening with Foliage, Form and Texture*, David & Charles PLC, Devon 2004

Rick Darke: *The American Woodland Garden*, Frances Lincoln, London 2000

Brian Hackett: *Planting Design*, McGraw-Hill, New York 1979

Richard Hansen, Friedrich Stahl: *Perennials and Their Garden Habitats*, Cambridge University Press, Cambridge 1993

Penelope Hobhouse: *Colour in Your Garden*, Collins, London 1985

Penelope Hobhouse: *Penelope Hobhouse's Garden Designs*, Frances Lincoln, London 2000

Gertrude Jekyll: *Getrude Jekyll's Colour Schemes for the Flower Garden*, Frances Lincoln, London 2006

Noël Kingsbury: *Gardens by Design*, Timber Press, Portland, OR 2005

Hans Loidl, Stefan Bernard: *Opening Spaces*, Birkhäuser Verlag, Basel 2003

Piet Oudolf, Noël Kingsbury: *Designing with Plants*, Conran Octopus, London 1999

Piet Oudolf, Noël Kingsbury: *Planting Design: Gardens in Time and Space*, Timber Press, Portland, OR 2005

Marco Valdivia, Patrick Taylor: *The Wirtz Gardens*, Exhibitions International, Leuven 2004

James Van Sweden, Wolfgang Oehme: *Bold Romantic Gardens: The New World Landscape of Oehme and Van Sweden*, HarperCollins Design International, New York 2003

Rosemary Verey: *Rosemary Verey's Making of a Garden*, Frances Lincoln, London 2006

图片鸣谢

Figures 1, 2, 12, 28, 32, 35, 40, 49, 51 (all), 52, 53, 56, 59, 61 (Sven-Ingvar Andersson Landschaftsarchitekt, Union Bank Kopenhagen), 67, 68, 71, 72, 76, 77, 79, 81, 82, 85, 86: Eva Zerjatke

Figure page 3, figures 4, 6 (Büro Kiefer, Schöneberger Südgelände, Berlin), 7 (DS Landschaftsarchitekten, Tilla-Durieux Park, Berlin), 8 (Stötzer Neher Berlin, VW AG glass factory, Dresden), 10, 14 (Kienast Vogt, Moabiter Werder, Berlin), 16, 29, 30, 33, 38, 39, 54, 55, 69, 75, 87 (Stötzer Neher Berlin, VW AG glass factory, Dresden): Hans-Jörg Wöhrle

Figures 5, 9, 11, 13, 15, 17 (in reference to H. Keller), 18, 19–21 (in reference to H. Schiller-Bütow) 22 (Sommerlad, Haase, Kuhli, shed in the Volkspark Bornstedter Feld, Potsdam), 23, 24, 25, 26, 27 (in reference to Günter Mader), 31, 34, 36 (in reference to Günter Mader), 36–37 (in reference to Hans Schiller-Bütow), 41, 42, 43 (in reference to Hans Schiller-Bütow), 44 (in reference to Herbert Keller), 45, 46, 47, 48, 50, 57, 58, 60, 62, 63, 64 (in reference to Brian Hackett), 65, 66, 70, 73, 74, 78, 80, 83, 84, 88: Regine Ellen Wöhrle

作者简介

雷吉娜·埃伦·韦尔勒（Regine Ellen Wöhrle）及汉斯－约尔格·韦尔勒（Hans-Jörg Wöhrle），工程硕士，执业景观建筑师，他们是柏林、斯图加特、施尔塔赫的三家 w＋p Landschaften 景观建筑事务所的所有人。